Reprinted from
Bibliographia Genetica XIX (1963): 317-404.

ISBN 978-94-017-6735-4   ISBN 978-94-017-6834-4 (eBook)
DOI 10.1007/978-94-017-6834-4

Bibliographia Genetica XIX (1963): 317–404.

THE INHERITANCE OF
PLUMAGE COLOUR IN THE COMMON DUCK

(*ANAS PLATYRHYNCHOS* LINNÉ)

F. M. Lancaster

National Institute of Poultry Husbandry,
Harper Adams Agricultural College,
Newport, Shropshire, England.
(*Received for publication April 17, 1961*)

CONTENTS

INTRODUCTION

There seems little doubt that the common domesticated duck originated from the Northern Mallard (*Anas platyrhynchos L.*). This is a surface-feeding duck of the Northern Hemisphere and includes four sedentary and two migratory subspecies. Of the latter, the Green-headed Mallard (*Anas platyrhynchos platyrhynchos* Peters) is the ancestor of the domesticated duck (DELACOUR, 1956). It is the commonest of the subspecies and has the largest geographical distribution. According to DELACOUR its breeding range extends over the Northern Hemisphere across North America, Europe and Asia, as far north as 70° in Norway down to the Mediterranean and North Africa. It winters as far south as Abyssinia, Arabia, India, China, Central Mexico and Florida.

The Mallard readily becomes tame in captivity and was probably the earliest bird to be domesticated. There have undoubtedly been many independent domestications throughout the ages. The Asiatic breeds, for example, have a different origin in time and place from the European breeds as demonstrated by their more vertical carriage. However, despite the duck's long period of domestication the mutant colour patterns in present day breeds are fewer in number and less complicated in their inheritance than in the fowl. This is probably a reflection of the fact that in the last century the fowl was more popular for exhibition than the duck.

The three chapters which follow form a survey of present knowledge of the inheritance of plumage colour in the common duck. Reviews of earlier work are supplemented by the results of experimental matings carried out at the National Institute of Poultry Husbandry during the past six years. These recent findings include two linkage groups, a triple allelic series, and two new loci.

The classification of the 11 loci (comprising 24 genes), which control plumage colour is purely arbitrary. The two loci described in the first section are the only multiple alleles known in this species. Further, they are the only groups of genes (excluding dilution genes) which produce small but generalized modifications to the wild-type plumage where the basic pattern is still apparent. The second section covers extended black and three dilution genes, two of which are sex-linked. In the final section the mode of inheritance of self-white and several types of white spotting is discussed.

Of the two known cases of linkage in the duck, (brown dilution – buff dilution), and (extended black – dominant bib), only the former has been investigated in this work for evidence of crossing over. These sex-linked loci were tested for linkage strength by backcrossing the double heterozygote to the hemizygote carrying both recessives. A limited examination of data concerning blue dilution and extended black in Chapter II and recessive white and extended black in Chapter III have failed to reveal any evidence of linkage.

In spite of JAAP and HOLLANDER's recommendation (1954) for the adoption of wild type symbolism (+) for all domesticated birds, it has not been universally accepted. Some authorities are reluctant to apply the wild type system to poultry since wild types are not easily determined in birds with polyphyletic or obscure origins such as the African goose and the common fowl. Although there is general agreement amongst biologists that the common domesticated duck has a monophyletic origin, with the Mallard as the sole ancestral species, the older method of notation is retained to avoid confusion. For those who prefer the wild type method, a table is included in the appendix which shows the wild type gene for each character. In addition to the genes for plumage colour described in this paper, the table also includes the yellow bill – white bill locus, the green egg – white egg locus, "crested" and "hereditary tremour", to complete the list of genes known in the duck.

To avoid repetition the wild-type pattern is described in detail at the beginning of the first section under the heading "the mallard pattern".

Facilities for this work were provided by the extensive collection of domesticated ducks maintained by the National Institute of Poultry Husbandry. This comprises 15 varieties of common duck and 5 varieties of muscovy. Since these were originally purchased from widely different sources they include most of the major mutant genes affecting plumage colour in Great Britain.

Wherever possible the breeds used in the experiments are specified, especially where several examples of the same genotype involving different breeds occur. Since confusion may arise between certain genes that have been named after varieties, such as "runner" and "mallard", and the varieties themselves, the convention followed throughout is to use capitals for breeds and varieties and small letters for gene descriptions.

The tables giving breed combinations for sex-linked crosses and the descriptions of phenotypes resulting from different gene combinations have been included because of their interest to the practical breeder.

References to the mode of inheritance in other species are restricted to cases where the situation is similar to that in the duck or where missing types can be found in other birds. In the section on blue dilution, the results of hybridization experiments are also included.

The three papers in this monograph were originally written for separate publication. It was later decided to include them in a single work since they cover the entire field of plumage colour genetics in the duck, a subject where the original reports are scattered widely throughout the literature.

CHAPTER I

MULTIPLE ALLELOMORPHS IN THE DUCK

Prior to these investigations, the only multiple allelic series known to control colour inheritance in *Anas platyrhynchos* was the "restricted – mallard – dusky" group described by JAAP (1934). Since the present work is partly concerned with interactions between the latter series and the new "dark phase – light phase – harlequin phase" group, (both of which cause autosomal variations in the pattern of the wild Mallard), it seems appropriate to summarize JAAP's observations on the "mallard" series in the same section.

THE RESTRICTED, MALLARD, DUSKY SERIES (M^R, M, m^d)

The "mallard" factor (M) allows full expression of the "wild type" pattern of the Mallard duck. It is dominant to the "dusky" factor (m^d) and recessive to the "restricted" factor (M^R). Each allele produces its own distinct pattern both in the day-old duckling and in the adult bird.

THE MALLARD PATTERN (M) (*Plates 1 and 2*) – The dorsal surface of the mallard duckling is olivaceous black in colour. Situated on the back and forming the corner points of a rectangle are four yellow spots. The sides of the head are also yellow and usually bear two dark ocular stripes running across the eye from the base of the bill to the dark dorsal area. The ocular stripes, however, are not so uniform as the dorsal spots and often vary in number and extent. The ventral surfaces of the upper neck and wings are yellow but the ventral surface of the body, although yellowish, has a black or slate undercolour which extends out to the surface and produces a greyish mottled appearance.

The adult mallard pattern is described under the standards for the Rouen and Brown Decoy (Grey Call) varieties and also in various ornithological works on the wild Mallard. The salient features are: –

Male: Head and neck rich iridescent green-black divided from the claret-coloured breast by a white neck ring. Back and rump – rich green-black running back from between the shoulders. Flank, side and belly – blue-grey, finely pencilled with black. Brilliant blue-green iridescent speculum edged with black and white bands on each wing. White ventral wing surfaces. Dorsal surface of wing and primary feathers – brownish grey.

Female: Head rich buff-brown finely streaked with brownish-black markings. The dark markings coalesce in certain areas to form a well-defined pattern. There is a dark line running from the base of the bill back through the eye and a dark patch on the crown which runs down the back of the neck. Wing specula are brilliant as in the male. Ventral wing surfaces white. The remainder of the plumage is a rich brown, each feather distinctly pencilled with black or very dark brown.

THE DUSKY PATTERN (m^d) (*Plates 1 and 2*) – In the day-old dusky the entire dorsal surface is olivaceous black as in the mallard. There is a shading off to dark olive-grey on the ventral side. The yellow dorsal spots and the ocular stripes of the mallard are absent. The whole of the dorsal surface and the head is uniformly dark and the ventral wing surfaces are pigmented.

The chief differences between the mallard and the dusky in the adult are: –

Male: There is usually a complete absence of white neck ring in the dusky. The claret colour of the breast of most dusky males is either missing or confined to a very small area at the base of the neck.

Female: The head is uniformly dark and lacks the eye stripe and cap of the mallard.

In both sexes the wing specula are obscured and the ventral wing surfaces are pigmented.

(The Khaki Campbell is a good example of a pure breed bearing the dusky pattern. – Author's note).

THE RESTRICTED PATTERN (M^R) (*Plates 1 and 2*) – The name "restricted" was applied to this factor because in the day-old duckling the area of dark pigment on the dorsal surface is confined to patches on the head and tail. The remainder of the dorsal area is dull yellow with a dark undercolour. The ventral surface is the same as that described for the mallard.

Adult restricted birds are very similar to individuals bearing the mallard pattern. In both sexes the chief differences occur on the wing front and bow where the restricted birds always show considerable areas of white on the dorsal surface. This is due to white lacing or tipping rather than completely white feathers. In the female the wing bow always appears much paler than in the mallard or dusky. This effect is independent of the presence of white and is due to a wider lacing of buff.

The above manifestations of the three alleles are summarized in Tables I and II.

THE DARK PHASE, LIGHT PHASE, HARLEQUIN PHASE GROUP (Li, li, li^h)

LIGHT PHASE. JAAP (1933b) described an autosomal mutation in semi-domesticated Mallards which he called "light phase" (li). The dominant allele of this factor, "dark phase" (Li), allows full expression of any one of the three "mallard" alleles which happens to be present; these have already been described. The light phase gene produces a further type of variation from the wild pattern. Its chief effect is to lighten the colour of the adult plumage, particularly in the female, and to give rise to greater sex dimorphism of plumage pattern. Although light phase has been described previously, (ROGERON, 1903; FINN, 1913), no controlled breeding experiments were carried out to determine its mode of inheritance until it was investigated by JAAP. His observations are summarized below.

Description of Ducklings. It is possible to distinguish between light and dark phase ducklings any time after 19 to 20 days of incubation. Light phase ducklings (*Plates 3 and 4*) can be recognised by the presence of white in the breast region. Although variable in extent these white areas are always present and can be used as an easy means of identification. Dark phase birds, unless modified by the "runner" factor (R), (JAAP, 1933a), always show complete pigmentation of the ventral surface of the body from the neck region to the tail.

The number of ocular stripes across the face can also be used as a guide in classification. In each case these dark stripes, which run through the eye, form a striking contrast with their pale yellow background. Their number depends on which of the two phases is present; light phase ducklings never have more than one unbroken dark stripe, whereas dark phase types, (except when carrying the dusky pattern – m^d), usually have two. However, very occasionally, dark phase ducklings do appear with only one stripe so that this characteristic can only be used in conjunction with others for an accurate classification.

In some cases it is very difficult to differentiate between light and dark phase in the presence of the sex-linked dilution factor (d) described by PUNNETT (1930; 1932), and WALTHER, HAUSCHILDT and PRÜFER (1932). In the day-old ducklings this factor tends to lighten the ventral surface of the dark phase and thus destroy the main basis for classification. When the two phases are associated with the dusky pattern (m^d) it is impossible to distinguish between them at day-old.

Differences at Maturity. The appearance of light phase at maturity depends upon which of the three "mallard" allelomorphs, M^R, M or m^d, occurs with it. In restricted and mallard phenotypes light phase produces a lighter colour in the juvenile, female and eclipse male plumages. This is brought about by a reduction in the size of the dark portion of each feather. In the male (breeding plumage), the most striking difference between dark and light phase in M^R and M birds is the increased area of claret brown on the breast. In light phase drakes the claret is extended along the sides and over the shoulders. JAAP states "It is interesting that a gene which produces a general reduction in colour in the female, should produce an increase in pigment in this region". A minor modification in the male, caused by the light phase factor, is the reduction of pigment on the anterior part of the back from black to dark grey.

In the dusky pattern, (dark phase), it will be remembered that one distinguishing feature in the male was the almost complete absence of claret in the breast region; the grey coloration extending forward to meet the iridescent greenish-black of the neck. In light phase duskies there is always some claret present in this area. A lighter tone occurs in the light phase dusky females than in the dark phase, on comparing the two phases, however, the difference is less than when either restricted or mallard is the background pattern.

All the possible interactions in the male between the two series in different combinations were summarised by JAAP in tabular form. (Table III.)

HARLEQUIN PHASE. HUNTER (1939) described a further variation from the wild pattern. The mutation appeared in a flock of wild Mallards which had been domesticated and inbred for about twenty-four years.

The day-olds were yellow with smoky-coloured down on their heads and tails. At maturity it was noted that much of the brown pigment of the female and the grey of the male was missing. The ducks were almost entirely white on the breast, had greyish heads and light-coloured wings and tails. The black markings of the male and the wing specula of both sexes were quite normal.

Three different matings were carried out to investigate the inheritance of this pattern. Unfortunately the number of ducklings obtained was small and classification was complicated by the presence of a dilution factor introduced from the Fawn Indian Runner. Nevertheless there was some evidence that this aberrant colour pattern behaved as a simple recessive to the wild type. These mutants have since been developed as a pure breed called the Light Mallard. (HUNTER, 1950).

In the F_2 generation from a Mallard drake $\times$ Indian Runner duck cross, PUNNETT (1932) obtained 4 light individuals from amongst 86 darker birds. The general ground colour was paler than in the Mallard, the breast was white, and the birds had a white collar. Although this pattern was undoubtedly a recessive character, the evidence available is insufficient to show whether it was JAAP's light phase or the pattern observed by HUNTER.

A British variety of duck, which appears to resemble the specimens of HUNTER, is the Welsh Harlequin. At day-old this variety has the pale orange-yellow down which is typical of white varieties such as the Aylesbury and the Pekin. However, it differs from the white breeds, by showing faint traces of smokiness on the head, tail and occasionally on the wings. The bill, legs and feet also bear traces of melanic pigment which gives these parts a slaty appearance never found in a white breed of duckling. On reaching maturity the birds resemble JAAP's light phase birds in pattern but are usually much paler in colour. The males also show a greater area of claret brown on the breast.

EXPERIMENTS

The similarity in appearance between the Welsh Harlequin and the 'light phase' duck prompted the idea that the two patterns might be caused by the same gene and that the differences were merely due to modifying factors present at other loci. To test this hypothesis several experimental matings were carried out at the National Institute of Poultry Husbandry in 1957 and 1958. The results are summarized in Table IV.

EXPERIMENTAL PROCEDURE

Crosses were made between birds of different genetic constitution as shown in Table IV. The colour markings of the F_1, F_2, backcross and other progeny were examined at three different stages: in the unhatched embryo, in the day-old duckling and at maturity.

Except for matings 1 and 2 (Table IV) most of the unhatched embryos which died after about the 20th day of incubation were examined and classified according to their patterns. As indicated by JAAP, the unhatched light phase duckling can be identified by the presence of large areas of white on the ventral surface. The dorsal surface of the light phase embryo, however, is mainly pigmented thus distinguishing it from the "Welsh Harlequin" type which is white throughout. The "dead-in-shell" from mating 2 could not be included because some pure recessive white (cc) ducklings segregated out and could only be distinguished from the Welsh Harlequin type in the hatched duckling. This explains the relatively small numbers obtained from this mating.

For economic reasons it was impossible to rear all the ducklings to maturity. Samples of each type from most matings were reared and in all cases adult classification was found to agree with that at day-old.

After recording the patterns of the day-old ducklings, they were identified to their pen, pattern and sex by means of toe punching. This was found to be more permanent than wing bands which easily became detached. By making several cuts on each web a large number of combinations could be obtained. All ducklings were vent sexed at day-old. In the case of matings involving the changing of drakes, an interim period of ten days was sufficient for the influence of the first male to disappear.

BREEDING RESULTS

F_1 GENERATION. In 1957 a pure-bred Rouen drake, with an unmodified wild-type plumage was mated with a Welsh Harlequin duck (mating 1). The colour of the 42 ducklings hatched was completely unexpected: 21 bore the wild-type dark phase pattern of the Rouen while the other 21 resembled the light phase ducklings described by JAAP. At maturity the two types agreed well with JAAP's descriptions of light and dark phase. Both patterns were distributed equally between the sexes.

It now seemed that if the light phase and the Welsh Harlequin patterns were identical, the latter must have been modified, in the F_1, by the introduction of other factors from the Rouen. Moreover, the Rouen drake used in the cross would appear to have been heterozygous for dark phase.

F_2 GENERATION. Two further matings (Nos. 2 and 3) were carried out in 1958. The former comprised intermating F_1 light phase phenotypes (li × li), and the latter intermating of F_1 dark phase phenotypes (Li × Li). Again the results were completely at variance with expectation. It was anticipated that the dark phase mating would produce a ratio of three dark to one light phase and that the light phase mating would be true breeding and yield all light phase ducklings. Instead, mating 2 (li × li) yielded 21 light phase ducklings and 4 Welsh Harlequin types, and mating 3 (Li × Li) produced 56 dark phase ducklings to 22 Welsh Harlequin types. Although the numbers obtained from the first mating were small, these figures suggest an interpretation based on a 3 : 1 ratio in both cases. (χ^2 for mating 3 = 0.43, P > 0.50). Thus, in addition to the known dark and light phases, a new Mendelian character showing clear-cut segregation appears to be involved. Further, there is good evidence that this new character forms the third member of a triple allelic series, with dark and light phase as the top dominant and dominant members respectively. Hereafter this new character will be referred to as the "harlequin phase" symbolised by the letters li^h.

Later in 1958 further matings were carried out to confirm the allelic relationship between Li, li and li[h]. The first of these consisted of mating together the two unlike F_1 heterozygotes (*i.e.* Lili[h] × lili[h]). This was carried out reciprocally (matings 4 and 5) and yielded the following phenotypes in the progeny: 57 Li, 26 li and 24 li[h]. This is a good approximation to the 2 : 1 : 1 expected ratio ($\chi^2 = 0.53$, $P = 0.8$).

BACKCROSSES. Both the F_1 generation heterozygotes were back-crossed to pure Welsh Harlequins (li[h]li[h]). These matings were again carried out reciprocally:

(*a*)   li[h]li[h] × lili[h]. (mating 6), lili[h] × li[h]li[h]. (mating 7)

(*b*)   li[h]li[h] × Lili[h]. (mating 8), Lili[h] × li[h]li[h]. (mating 9)

The following phenotypic results were obtained:

Group (*a*) – 73 li, 78 li[h].

Group (*b*) – 33 Li, 29 li[h].

Both results show a close approximation to the expected 1 : 1 ratio. ($\chi^2 = 0.17$ and 0.26, respectively; $P > 0.50$ in both cases).

OTHER MATINGS. The first mating (no. 1) was repeated in 1958, as originally planned, using a homozygous dark phase Rouen drake with Welsh Harlequin females – LiLi × li[h]li[h], (mating 10). This yielded 57 ducklings which were all dark phase.

It will be seen later, from a detailed description of the three phases, that the amount of pigment present (except for the claret on the male's breast, the green-black of the male's head and neck and the wing specula), becomes progressively reduced and replaced by white, in passing through the series from the top dominant to the bottom recessive. The ultimate and logical limit of this progression, on the female side at least, would seem to be pure white. It was decided therefore to find out whether or not recessive white (cc) was the true bottom recessive and thus formed a fourth member of the series. A Welsh Harlequin drake was mated with White Campbell ducks: 39 ducklings were obtained, all coloured and all dark phase, showing that the recessive white factor of the White Campbell is situated at a different locus from the other three phases (mating 15).

To determine whether or not the harlequin phase was true breeding and did not mask any other recessives in the present strains, the results from the pure Welsh Harlequin breeding pens were noted for three years. 74 ducklings were hatched over this period, none of which deviated very far from the normal harlequin phase pattern, (matings 11, 12 and 13). In addition two F_2 generation harlequin phase birds were mated together (mating 68). On this occasion 17 ducklings were hatched – all harlequin phase.

The final mating (no. 14) is not a true back-cross, and is not included under that heading, because the original Rouen parent was shown to be heterozygous for dark phase (Lili). It was included to provide further evidence of the autosomal nature of the series. An F_1 heterozygous male bearing the two recessive members of the series (lili^h) was mated with pure dark phase females. If sex linkage had been present half of the females would have appeared light phase and half harlequin phase, whereas all the males would have been dark phase. Instead, all the ducklings, male and female, (26), were dark phase demonstrating autosomal inheritance and confirming JAAP's observations.

EVIDENCE FOR ALLELISM OF Li–li–li^h. The foregoing data have shown that light phase (li) behaves as a simple dominant to harlequin phase (li^h). Regarding the position of the dark phase pattern (Li), the results of the first experiment (mating 1) could be interpreted in three ways: the dominant allelism of dark phase over the other two types; close linkage between light phase and a dilution factor which changes light phase to harlequin phase; or the presence of a separate pair of factors at a different locus – the dominant member of which would be epistatic to the patterns at the li–li^h locus.

Each of the above hypotheses will be examined further by comparing the expected with the observed results in matings 8 and 9. Table V sets out these results according to the multiple allelic hypothesis, Table VI demonstrates the behaviour of pattern genes which are linked with a recessive dilution factor, and Table VII shows the effect of epistasis.

The third hypothesis, relying on epistasis between two gene pairs, is untenable since the results obtained are not in agreement with expectation. It can also be seen that the expected results in Tables V and VI are only identical when linkage is very close. If any of the cross-overs were to appear, liH would be light phase and Lih would presumably represent a new fourth phenotype. Since neither have been observed to date it seems preferable to adhere to the allelic hypothesis until further evidence becomes available.

DESCRIPTION AND COMPARISON OF THE THREE ALLELES

DARK PHASE – LIGHT PHASE. The dark phase – light phase differences depend very largely on which member of the $M^R - M - m^d$ series is present. These will be examined separately.

The mallard combinations (*Plates 3 and 4*) – The differences between light and dark phase ducklings bearing the mallard pattern (M) have been described already in the review of JAAP's work.

At day-old the light phase ducklings have white areas on the breast and a single ocular stripe compared with a pigmented breast and double ocular stripes of the dark phase. A further feature of light phase, not mentioned by the above author, but which was observed in the stock of ducklings at this Institute is that the four dorsal spots of the mallard pattern, which incidentally can be found in most wild species of ducks, become elongated and exaggerated in the presence of li.

In the adult light phase mallards JAAP noticed a general lightening in colour of the female due to a restriction of the dark part of each feather. In addition practically the whole of the lower breast and abdomen is pure white as opposed to the normal brownish abdomen of dark phase. Other areas are also paler due to unbroken patches of white amongst the coloured feathers. In both phases, male and female, the ventral surfaces of the wings are white.

In the adult light phase drake the claret area of the breast is extended down the sides and over the shoulders. A greater part of the general vermiculated grey body colour of the male is replaced by white and the white neck ring is wider and more distinct than in dark phase. With the exception of the anterior dorsal region, which is dark grey in the presence of li, the sex dimorphic black areas are similar in both phases.

The dusky combinations. As stated earlier the dusky pattern (m^d) prevented Jaap from classifying the ducklings at day-old. When examination was confined to down colour only this was confirmed in our own stock where no detectable differences could be seen between dark and light phases carrying the dusky pattern. A marked difference was noted, however, in the colour of the legs and feet of the two phases. In these areas and sometimes, but not always, in the bill the colour was much paler in the light phase birds.

At maturity the light phase dusky ducklings in some respects resemble the dark phase mallard pattern since the males often have a white neck ring and claret breast and both sexes sometimes have white ventral wing surfaces. In Jaap's strains of ducklings modifying genes prevented the white neck ring and white ventral wing surfaces from developing but not the claret breast. The chief differences between M Li and m^d li are: –

(1) The general body colour in the female and the colour of the abdomen in the male are much paler in the light phase dusky than in the dark phase mallard. There is no replacement with white, however, as in the light phase mallard and light phase restricted phenotypes.

(2) The wing specula of m^d li types are obscured as in the dark phase dusky.

(3) The females lack the eye stripe present in the mallard pattern.

The restricted combinations. Light phase restricted (M^R) ducklings have the same basic pattern at day-old as dark phase in that the dark pigment is restricted to the head and tail. The contrasting paler areas, however, are much paler than in the dark phase birds and the ventral body surface of the duckling is white or pale yellow as in light phase mallards. At the surface the pale areas on the back are pale yellow and it is possible on superficial examination for some confusion to arise between light phase restricted and harlequin phase ducklings. On closer examination, however, it can be seen that the undercolour of the pale down of the light phase restricted is dark whereas in the harlequin phase the pale colour goes down to the base of the down feathers. Furthermore, the dark areas on the head and tail tend to be more extensive in the restricted light phase than in the harlequin phase.

At maturity much of the coloured plumage of the dark phase restricted is paler and partially replaced by white in the light phase. In general appearance M^R li ducklings resemble light phase mallards (M li). The main difference lies in the colour of the dorsal surface of the wing fronts which is pigmented in the M li combination but mainly pure white in M^R li. The amount of white is much greater and more continuous than in the dark phase restricted duck and superficially resembles the wing front of the harlequin phase, *vide infra*. Again closer examination of li M^R reveals that the undercolour in this area is pigmented whereas the feathers and down are pure white through to the skin in li^h.

It was pointed out under the review of Jaap's work that in some instances it was very difficult to distinguish between dark and light phase in the presence of brown dilution (d). This problem has not been encountered in any of the present experiments in either the unhatched embryo or the day-old, even though many progeny were homozygous or hemizygous for d. It must be presumed, therefore, that Jaap's diluted ducklings were further lightened by other modifiers to produce a paler type of duckling than any found at this centre.

LIGHT PHASE – HARLEQUIN PHASE (*Plates 3 and 4*) – The day-old harlequin phase duckling is completely yellow except for a faint smokiness on the head and tail, whereas practically the whole of the dorsal surface of the light phase duckling is pigmented. Whenever harlequin phase ducklings are homozygous or hemizygous for brown dilution (d) the smokiness is less apparent than when D is present.

Unlike light phase the presence of li^h prevented recognition of the three "mallard" alleles and this epistasis was found to persist to maturity. The difficulty arose through li^h obscuring the phenotypic expression of M^R, M and m^d in the main areas used for classification, e.g. M^R, M and m^d harlequin phase ducklings always have a white ventral wing surface whereas in dark phase the same area in the dusky is pigmented. Also, due to the lack of pigment present, the eye stripe of M birds is difficult to see, the day-old restricted pattern is obscured for the same reason. At maturity $li^h m^d$ drakes have a white neck ring which is absent in Li m^d birds.

The adult females of li^h are generally lighter in colour due to paler individual feathers and to larger solid areas of white. The coloured feathers appear paler because of a further reduction in the amount of black on each feather, a dilution of the brown portion and replacement of part of the brown area by white. The same process occurs to a lesser extent when dark phase is converted to light phase. One accurate method of classifying the adults (in both sexes) is to examine the dorsal wing surfaces. In light phase practically the whole of the dorsal wing area including the primary feathers is pigmented but in harlequin phase much of this area is white. This non-pigmented area on the wing web is caused by large areas of pure white. It should not be confused with the white tipping or lacing found in the light and dark phase restricted patterns. In addition the proximal half of each primary feather is white in the harlequin phase. At the National Institute of Poultry Husbandry the ventral surfaces (wings, breast and abdomen) of both li and li^h are white even in "dusky" type birds.

In the drakes of both phases the anterior part of the back is grey, but the grey is much lighter in harlequin phase. The males can often be classified on the amount of claret on the breast. The distribution across the shoulders is the same in both types but it extends further down the breast in li^h than in li.

The effect of brown dilution (d) on the two plumage patterns li and li^h is to obscure the wing speculum. Other effects of d, found on closer examination, are as follows: –

(1) A browner tinge in the grey of the drake's back; (2) The distal half in li^h and the whole of the primaries in li are paler when d is present; (3) The upper tail coverts and other black areas around the drake's tail are diluted from black to dark chocolate by d; (4) The greenish black appearance of the head and neck of the diluted (d) males is slightly less brilliant than in the non-diluted (D) types; (5) In the females of both phases the black portion of each contour feather is diluted to khaki brown.

The different phenotypic expressions of the three phases in conjunction with the "mallard" pattern (M) are summarized in Tables VIII and IX.

THE SILVER APPLEYARD DUCK. The Silver Appleyard duck is an ornamental breed of British origin. There are two types bearing the same pattern – a standard variety and a bantamized version. During the investigations just described it was assumed from written communications, that the pattern of this variety was the result of the harlequin phase gene.

In 1961 some mature Silver Appleyards were purchased and mated in different combinations with dark phase, light phase and harlequin phase ducks. About 150 ducklings were hatched from these matings. They showed conclusively that the pattern of the Silver Appleyard was due to a combination of the light phase and restricted patterns (li M^R). Later, when pure Appleyards were hatched it was found that they were typical of the li M^R phenotype at day-old which explains the earlier confusion with harlequin phase.

DISCUSSION

The present experiments show that dark, light and harlequin phases form a triple allelic series. The light phase pattern has not been reported before in Great Britain and indeed it would probably not have come to light in these experiments if the Rouen drake chosen for crossing with a Welsh Harlequin duck, in the first mating, had been homozygous for dark phase. The frequency of li in the N.I.P.H. Rouens must be very low and possibly the result of a recent mutation, because even though the birds have a high coefficient of inbreeding due to continuous sib and half-sib matings for five generations, no aberrant types have been observed before. It was, therefore, fortuitious that the three phases were brought together in the same experiment.

The light and harlequin phases produce a progressive reduction in the pigment of the wild type dark phase. This is a further indication that they form a true multiple allelic series and are not the result of closely linked factors. The only exceptions to the above are in the wing specula of both sexes, the green-black of the male's head and neck and the claret-brown of the male's breast. The claret-brown does in fact become extended in progressing through the series. JAAP's observations, therefore, as to the semi-pleiotropic nature of the light phase is equally applicable when referring to the harlequin phase.

SUMMARY

(1) Previous observations on the "restricted-mallard-dusky" triple allelic series (M^R, M, m^d) in the common duck are summarized. The two mutant genes produce minor modifications to the "wild-type" pattern.

(2) The mallard pattern can be seen in the Rouen, the Mallard and the Brown Decoy breeds; the dusky in the Khaki Campbell; and the restricted in the Silver Appleyard.

(3) Investigations into the relationship between certain genes responsible for further minor modifications to the wild-type pattern are described.

(4) Dark phase (Li), light phase (li) and harlequin phase (li^h) have been shown to form a triple allelic series in the above order of dominance.

(5) Dark phase is typically represented in the Mallard and Rouen varieties. Harlequin phase is found in the Welsh Harlequin and (apparently) in Hunter's Light Mallards. Light phase is present in the Silver Appleyard in combination with the restricted pattern.

(6) Light and harlequin phases produce a progressive reduction of pigment in all areas except the wing specula, green-black head and neck and claret breast. In the latter area the pigment is actually increased in the direction of the recessive.

(7) Recessive white (cc) has been shown to occur at a different locus and is not a member of the series.

(8) The effects of brown dilution (d) on the two mutant phases are described.

(9) Wherever possible interactions between the two multiple series are discussed.

TABLE I

APPEARANCE AT DAY-OLD (M, m^d AND M^R)

Type	Dorsal Spots	Ocular Head Stripes	Dark Down Pattern
mallard (M)	Present	Present	Fully Extended
dusky (m^d)	Absent	Absent	Fully Extended
restricted (M^R)	Obscured	Obscured	Restricted

TABLE II

ADULT APPEARANCE (M, m^d AND M^R)

Type	Ventral Wing Surface	Ocular Head Stripe (♀♀)	Dorsal Surface of Wing, Front and Bow	Wing Speculum	Claret breast and White Neck Ring (♂♂)
mallard (M)	White	Present	Pigmented	Present	Present
dusky (m^d)	Pigmented	Absent	Pigmented	Obscured	Absent
restricted (M^R)	White	Present	White	Present	Present

TABLE III

PHENOTYPIC VARIATION OF THE DIMORPHIC PLUMAGE AREAS OF MALES, AS COMPARED WITH THE GENERALLY ACCEPTED WILD MALLARD TYPE, MLi

(JAAP, 1933b)

COMBINATIONS:

M^R and
- Li – white spotted feathers on the dorsal wing surface.
- li – white spotted feathers on the dorsal wing surface; claret-brown along the sides of the breast.

M and
- Li – "Wild Type".
- li – claret-brown along the sides of the breast.

m^d and
- Li – absence of claret-brown on the breast; absence of sexual white neck ring; stippled grey ventral wing surface.
- li – absence of sexual white neck ring; stippled grey ventral wing surface.

TABLE IV

TRIPLE ALLELIC SERIES. (Li–li–li^h). EXPERIMENTAL RESULTS

Mating No.	Sire	Dam	Genotypes	Phenotypes of Progeny		
				Li	li	li^h
1.	Rouen	Welsh Harlequin	$Lili \times li^h li^h$	21	21	—
			Expected	21	21	—
2.	F_1 light phase	F_1 light phase	$li li^h \times li li^h$	—	21	4
			Expected	—	18.75	6.25
3.	F_1 dark phase	F_1 dark phase	$Li li^h \times Li li^h$	56	—	22
			Expected	58.5	—	19.5
4.	F_1 light phase	F_1 dark phase	$li li^h \times Li li^h$	15	7	9
5.	F_1 dark phase	F_1 light phase	$Li li^h \times li li^h$	42	19	15
			Total	57	26	24
			Expected	53.5	26.75	26.75
6.	Welsh Harlequin	F_1 light phase	$li^h li^h \times li li^h$	—	18	12
7.	F_1 light phase	Welsh Harlequin	$li li^h \times li^h li^h$	—	55	66
			Total	—	73	78
			Expected	—	75.5	75.5
8.	Welsh Harlequin	F_1 dark phase	$li^h li^h \times Li li^h$	23	—	20
9.	F_1 dark phase	Welsh Harlequin	$Li li^h \times li^h li^h$	10	—	9
			Total	33	—	29
			Expected	31	—	31
10.	Rouen	Welsh Harlequin	$LiLi \times li^h li^h$	57	—	—
			Expected	57	—	—
11, 12, & 13.	Welsh Harlequin	Welsh Harlequin	$li^h li^h \times li^h li^h$	—	—	74
68.	F_2 harlequin phase	F_2 harlequin phase	$li^h li^h \times li^h li^h$	—	—	17
			Total	—	—	91
			Expected	—	—	91
14.	F_1 light phase	Rouen	$li li^h \times LiLi$	26	—	—
			Expected	26	—	—
15.	Welsh Harlequin	White Campbell	$li^h li^h CC \times LiLicc$	39	—	—
			Expected	39	—	—

TABLE V

AN INTERPRETATION BASED ON THE HYPOTHESIS OF 3 ALLELIC GENES IN THE
FOLLOWING ORDER OF DOMINANCE. – Li, li, li^h

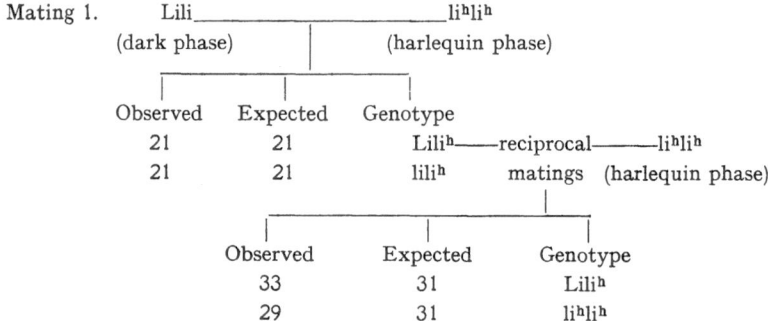

TABLE VI

SHOWING THE INHERITANCE OF THE THREE COLOUR PATTERNS WHERE LIGHT
PHASE (li) IS CLOSELY LINKED WITH A DILUTION FACTOR (h)

Li — dark phase,   H — non-diluted,
li — light phase,   h — diluted.

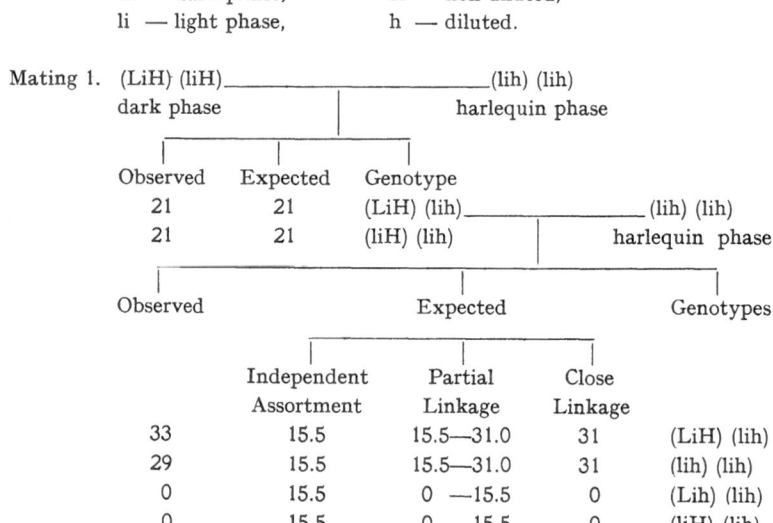

TABLE VII

TO SHOW THAT A SEPARATE FACTOR PAIR WITH EPISTASIS OF THE DOMINANT DOES
NOT ACCOUNT FOR THE RESULTS OBTAINED

D – dark phase,   li – light phase
d – non-dark phase,   li^h – harlequin phase.

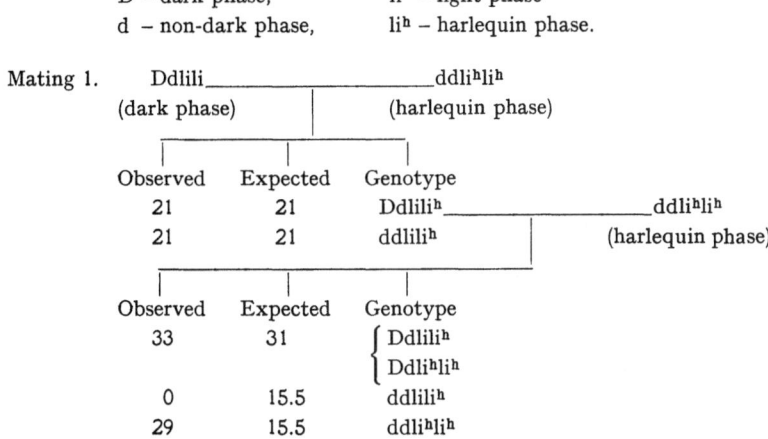

Mating 1.   Ddlili_____ddli^{h}li^h
(dark phase)   (harlequin phase)

Observed	Expected	Genotype
21	21	Ddlili^h_____ddli^{h}li^h
21	21	ddlili^h (harlequin phase)

Observed	Expected	Genotype
33	31	{ Ddlili^h
		{ Ddli^{h}li^h
0	15.5	ddlili^h
29	15.5	ddli^{h}li^h

TABLE VIII

DAY-OLD CHARACTERISTICS OF Li, li AND li^h

Type	Dorsal Body Surface	Ventral Body Surface	Dorsal Spots	Ocular Stripes
Li	Pigmented	Pigmented	Normal	Usually two
li	Pigmented	White	Elongated	One
li^h	White	White	Obscured	Obscured

TABLE IX

EXPRESSION OF Li, li AND li^h AT MATURITY

Type	Distribution of Claret on Breast of Male	Extension of Body Pigment (grey in male and brown in female)	Dorsal Wing Surface	Ventral Body Surface (♀♀)	White Neck Ring (♂♂)
Li	Normal (breast only)	Normal	Pigmented	Pigmented	Normal
li	Extended on to shoulders and sides	Partially replaced by white	Pigmented	White	Enlarged
li^h	Further extended (shoulders and sides as above and posterior extension on breast)	Mainly replaced by white	Mainly White	White	Enlarged

CHAPTER II

SEX-LINKAGE AND THE INHERITANCE
OF DILUTION FACTORS

The two aspects of colour inheritance mentioned in the title of this section are treated simultaneously because two of the three dilution factors to be discussed are carried on the sex chromosome. Two of these factors have already been observed and analyzed by other workers, (PUNNETT, 1930; WALTHER, HAUSCHILDT and PRUFER, 1932; JAAP and MILBY, 1944), but as the present experiments show, they have a wider application than was originally reported. The third factor, which is sex-linked, has not been described before and came to light in the course of other investigations.

It is not the purpose of this paper to examine the chemical, physiological and histological reasons for the dilution of pigment. This subject has been given adequate consideration elsewhere, (LLOYD-JONES, 1915; LIPPINCOTT, 1918 and 1923; MASON, 1923 and 1927; DU SHANE, 1944; FOX, 1955). The present report is confined to the genetics and phenotypic expression of the above factors and their interaction effects.

Before considering dilution factors which modify the intensity of pigment it is first necessary to examine the basic background patterns on which they operate.

EXTENDED BLACK (E) AND NON-EXTENDED BLACK (e)

In the common duck fully extended black is a simple autosomal dominant to non-extended black (PHILLIPS, 1915; JAAP and MILBY, 1944). Breeding results at the National Institute of Poultry Husbandry have also confirmed this (Table X). When homozygous, the recessive factor (e) allows full expression of the "wild pattern" of the Mallard duck or any of its variations caused by mutations at the M and Li loci (reference JAAP, 1933b, 1934; and Chapter I of this paper). In these patterns black is restricted to the sex dimorphic black areas of the male but in the female the restriction varies according to the basic pattern present. The dominant allele (E) causes solid black pigment to be laid down in all areas except those which are controlled by genes for white spotting. This latter aspect will be examined in Chapter III of this work. It is sometimes difficult to distinguish between black (EE??) and "dusky mallard" (eem^{d}m^d) ducklings at day-old. The only reliable way is to examine the legs and feet where the melanic pigment is much more intense in the black individual. Dusky mallards also carry olivebrown "guard hairs" in the down which are absent in the black.

For the purpose of the following discussions, unless otherwise stated, it can be assumed that when e is referred to in the homozygous state its phenotype is the "wild type" pattern ("mallard" – M, "dark phase" – Li).

BLUE DILUTION

Blue dilution of black pigment can be found in most domesticated birds. Its mode of inheritance shows considerable variation not only between but also within species. However, in birds normally covered by the term "poultry", with a few exceptions, the blue dilution factor usually behaves as an imperfect dominant and has a greater effect in the homozygous than in the heterozygous state (Table XI).

The existence and mode of inheritance of blue dilution in the duck has been known for many years. WRIGHT (1902) states that about 1860 there existed in Lancashire a local breed of blue duck which could be produced by crossing "white" with either black or any dark breed like the Rouen. LAMON and SLOCUM (1922) point out that the Blue Swedish duck behaves very much like the Blue Andalusian fowl with regard to colour inheritance. GHIGI and TAIBEL (1927) also show that the common blue duck is heterozygous for blue dilution. Finally JAAP and MILBY (1944) working with Blue Swedish ducks, provide evidence that the colour of this breed is due to the heterozygous expression of a dominant autosomal gene.

It was necessary in the early stages of this work to establish whether the blue dilution of the Blue Orpington was the same type as that of the Blue Swedish on which most of the original work was done. The possibility of another form of blue dilution in the Blue

Orpington could not be ignored since other species of domesticated poultry are known to carry sex-linked and recessive forms in addition to the dominant autosomal type (Table XI). Table XII shows the results obtained from different matings involving the Blue Orpington in all its forms. It provides sufficient evidence to show that the same type of blue dilution is involved in both varieties.

The symbol originally assigned to the blue dilution factor by JAAP and MILBY (*loc. cit.*) was G for both ducks and turkeys. This has since been changed by JAAP and HOLLANDER (1954) to Bl for ducks and D for turkeys. Since D is the allelomorph of brown dilution in the duck it would seem preferable to retain the older symbol G for both species to be more consistent and to avoid confusion.

G is primarily a diluter of black pigment and has little effect on other colours. The final result, therefore, depends on which background pattern it is associated with.

BLUE DILUTION OF EXTENDED BLACK. In the presence of E blue dilution produces a uniform blue colour throughout the bird (except for a slight darkening in the sex dimorphic areas of the male and black flecking in the heterozygote). The shade of blue depends on whether G is homozygous or heterozygous (*Plate 5*). In the adult homozygous female the plumage colour is practically pure white and can often only be distinguished from a recessive white duck by its dark bill and eye colour which is not affected by blue dilution. The male homozygote does possess slightly more pigment and has a very pale grey appearance (*Plate 6*). The heterozygous birds of both sexes are much darker and, as stated above, always exhibit solid areas of black flecking against the blue.

SEREBROVSKY (1926) suggested that this flecking in fowls was due to repeated loss of a chromosome bearing the dominant dilution factor in the heterozygote. However, HOLLANDER (1944) points out that variegation of this type could equally well be explained by the fact that Bl is a highly mutable gene. SEREBROVSKY states that flecking can also be found in homozygous blue fowls; in this case the flecking is blue among dirty white feathers as opposed to black amongst blue in the heterozygote. Blue splashing in the homozygote has not been observed in either ducks or turkeys at this Institute. In turkeys, ROBERTSON (1925) also found that variegation was only evident in birds which were heterozygous for blue dilution and he suggested somatic non-disjunction of chromosomes as a possible explanation.

At day-old, homozygous blue in combination with E produces a very pale cream-coloured down whilst the bill again remains dark lead-blue and the legs and feet are slate. In the day-old Blue Orpington the colour patterns GG and Gg are quite distinct in shade (*Plate 7*). In the Blue Swedish, however, the variation in duckling colour at hatching is sometimes so great that there is no sharp line of demarcation between the two types (JAAP and MILBY, 1944).

Due to the presence of modifying factors in some breeds and strains there appear to be two different types of black in the duck: –

(1) That of the Black Orpington which is a dull muddy black except for the head, neck and back of the male which have a greenish sheen, and

(2) That of the Black East Indian and Cayuga varieties where the birds carry this irridescent green sheen all over the body and the black is more intense. The male is more brilliant than the female.

When the former type of black is diluted with G it tends to show traces of reddish-brown pigment down the sides and flanks and in extreme cases over the shoulders and wing coverts. This fault is more pronounced in the male. If the second type of black is used the resulting blue shows no trace of "rustiness" in either sex. There is no rustiness present in either the day-old or growing duckling of either type, it is only apparent at maturity. Since these rusty birds are pure for extended black (E) and non-brown dilution (D) it is suggested that the brown pigment is due to the presence of genetic modifiers at other loci.

In the present experiments no evidence of linkage was found between the two autosomal factors E and G (Table XIII).

BLUE DILUTION OF NON-EXTENDED BLACK. In the non-extended (wild type) black (e) the black pigment is restricted to certain areas of the body as stated previously. It is only in these areas that blue dilution is able to exert its full effect. Other colours, particularly brown, are only very slightly affected by G. When blue dilution is introduced into the Rouen or Mallard varieties without E a peculiar effect is produced where, except for the black areas the plumage colour remains practically unchanged. The black areas, are diluted to blue, the shade of which varies according to whether G is present in a single or a double dose.

At day-old these "Blue Mallards" are mainly blue (*Plate 8g*), but sometimes vaguely show the wild pattern of the Mallard duck. If they carry the dusky pattern (m^d) they are indistinguishable in colour from the ordinary blue (EEGg) duckling. They can still sometimes be classified, however, by observing whether the dominant white bib (S), which is associated with the Blue Orpington and Blue Swedish varieties, is present or absent. (Compare top row with bottom row in *Plate 8*). As Chapter III of this series of papers will show "bib" is closely linked with extended black (E) and is, therefore, nearly always transmitted with it. At day-old "homozygous" Blue Mallards (GGee) cannot be differentiated from "homozygous" Blues (GGEE).

There is no pure-bred variety of duck bearing this combination of factors.

BROWN DILUTION

Brown dilution in the duck was first reported by PUNNETT (1930 and 1932) and by WALTHER, HAUSCHILDT and PRÜFER (1932). The latter investigations were carried out independently of PUNNETT but at about the same time. Both authorities came to the same conclusion that this factor was a sex-linked recessive.

Brown dilution modifies the wild pattern in much the same way as blue dilution. It changes the intensity of pigment in certain areas but does not affect the basic pattern. In other domesticated birds the nearest counterpart to brown dilution of the duck can be found in the turkey where the colour of the brown variety is also caused by a recessive sex-linked factor – e (ASMUNDSON, 1945). In the turkey brown dilution changes black pigment to chocolate brown without affecting the wild pattern of the Bronze.

In the experiments of PUNNETT (1930, 1932) and WALTHER *et al.* (1932), the breeds of duck used were all of the non-extended black (e) type. However, as noted by JAAP and MILBY (1944) d is primarily a diluter of black pigment and, therefore, produces a different result in the presence of E. The different effects of brown dilution on the two basic patterns (E and e) will now be examined separately.

BROWN DILUTION OF NON-EXTENDED BLACK (*Plates 8f and 9c*) – In PUNNETT's experiments Mallards, Fawn Indian Runners and

Khaki Campbells were used. WALTHER *et al.* used Khaki Campbells and White Indian Runners. All these varieties are homozygous for e: The Mallard duck bears the non-diluted (D) "wild type" plumage whereas the Fawn Runner and the Khaki Campbell, although modified by variations at the M locus, carry brown dilution (d). This has the effect of diluting the sex dimorphic black areas on the head, neck, back and rump of the male to dark chocolate brown. All black lacing on the female, black peppering on the male's flanks and other black areas are modified to the same colour. The light brown areas of the "mallard" pattern, which occur chiefly in the female, remain unchanged. The brilliant wing specula of both sexes are obscured by d. The black and blue areas in the speculum are changed to varying degrees of brown; only the white stripes are unaffected.

The colours of the day-olds of both types were described by PUNNETT in his first paper (1930) by reference to RIDGEWAY's Colour Standards (1912). The dark non-diluted duckling colour was almost equivalent to RIDGEWAY's "dark olive" and the diluted duckling corresponded with his "light brownish-olive".

Most white ducks are of the non-diluted, non-extended black type (D(D)ee). Since the epistatic white factor (c) is an autosomal recessive they behave in the same way as the Mallard duck in a sex-linked cross.

At the National Institute of Poultry Husbandry the same results were obtained by using the following breeds: – Welsh Harlequin, Khaki Campbell, Rouen and White Campbell. The first two carry brown dilution (d), and the second two have the dominant allele (D). The recessive "harlequin phase" (li^h) of the Welsh Harlequin did not appear in any F_1 offspring because the dominant "dark phase" was introduced from the Rouen and White Campbell. The matings were carried out reciprocally as shown in Table XVIII.

BROWN DILUTION OF EXTENDED BLACK (*Plates 8b and 10b*) – Brown dilution of extended black (d(d)EE) produces a fairly uniform dark chocolate appearance in both sexes. The male does, however, show variation in the sex-dimorphic areas which are slightly darker and more glossy than in the female. Pure breeds possessing this combination are the Chocolate Orpington and the Chocolate Runner (both very rare). Chocolate females can also be produced in a sexlinked cross using a Khaki Campbell male with a black female. Table XVIII shows the results of several such matings where Black Orpington and Black Cayugas were used on the female side. Four of these matings resulted in the following progeny:

33 black males, 32 chocolate females.

The same effect can be produced by crossing a chocolate male with either a black or a Rouen female.

To distinguish between a Khaki Campbell (diluted dusky – $d(d)eem^dm^d$) and a chocolate duckling ($d(d)EE??$) at day-old one should examine the legs and feet which are darker in the chocolate. The down of the chocolate is also darker due to the presence of fewer light olive-brown "guard hairs". As stated earlier, dominant white bib is usually associated with E due to close linkage and can be used as a further guide in classification at day-old.

DILUTION COMBINATIONS

Up to the present stage blue and brown dilutions have been examined separately. They can, however, occur together in the same individual either in the presence of e or E. This produces a further series of colour variations which will be described below.

BLUE AND BROWN DILUTION OF NON-EXTENDED BLACK (*Plates 7b, 7c, 8h, 9a and 9b*) – JAAP and MILBY (1944) first showed that the combined effects of blue and brown dilution on non-extended black ($GGd(d)ee$) produced the well-known buff colour in ducks. In Great Britain this colour is seen typically in the Buff Orpington which was formerly a popular dual purpose variety. Experimental matings, carried out at this centre, have shown that two definite shades of this colour variety exist as in the Blue Orpington (Table XIV). This variation is due to the different phenotypic effects of blue dilution according to whether it is homozygous or heterozygous. As shown earlier homozygous blue has a greater dilution effect and produces a paler type of bird than heterozygous blue. Unfortunately the British Standards for these two Orpington varieties have been fixed to coincide with the darker of the two shades in both cases. It is inevitable, therefore, that there will be a 50% wastage of segregates in each generation due to the heterozygous nature of the standard-bred parents. This probably accounts for the decline in popularity of these breeds.

The recessive segregates (d(d)ggee) from the standard Buff Orpington (d(d)Ggee) resemble the Khaki Campbell except for certain
differences due to a further dilution factor which will be discussed
later. The mature "homozygous" Buff Orpington females are of a
much paler buff colour than the "heterozygous" ones especially on
the wings. In the male the greatest differences are found in the sex-
dimorphic black regions of the "wild pattern" (head, neck and lower
back). In the "homozygous" buff male these areas are a pale bluish-
brown colour, almost lilac, but in the "heterozygous" (standard)
male they are dark tan.

At day-old the differences between the three types resulting from
standard Buff Orpingtons mated inter se are not outstanding. The
most accurate method of classification is by examination of the
legs and feet. These are pink in the "homozygous" buff, they show a
slight trace of pigment in the "heterozygote" but are dark in the
recessive segregates (gg). There is gradual darkening of pigment
in the down from the homozygous dominant to the homozygous
recessive. The down colour of the latter is similar to that of the pure
Khaki Campbell, the bill, legs and feet, however, are not so dark.
The M^R, M and m^d series cannot be differentiated at day-old in the
presence of GGd(d)ee; even with "heterozygous" buff, Ggd(d)ee,
classification is still very difficult. "Homozygous" buffs at day-old
cannot be distinguished from other colour combinations carrying
the homozygous blue dilution factor. These are: pale blues (GGD(D)
EE), pale blue mallards (GGD(D)ee) and pale lilacs (GGd(d)EE).

BLUE AND BROWN DILUTION OF EXTENDED BLACK (*Plates 8d and
10a*) – Blue and brown dilution working together on extended black
produce a reddish or brownish shade of blue which can best be
described as "lilac". This colour is only expressed fully when blue
dilution is heterozygous. When G is homozygous a lilac-splashed
white effect is produced corresponding to the blue-splashed white
of the "homozygous" Blue Orpington. In the dark lilac (Gg) the
black flecking of the heterozygous blue is replaced by chocolate
flecking. The sex-dimorphic head, neck and back regions are darker
in the male. The male "homozygotes" also show more pigment than
the females.

At day-old the lilac duckling closely resembles the buff in down
colour but the lilac usually shows darker pigmentation of the legs
and feet and has a white bib.

There is no pure breed of duck corresponding to this dark lilac colour. The objection to such a colour, however, is that it suffers from the same disadvantage as the heterozygous blue and the dark buff varieties where there is a 50% wastage of "homozygous" colours (chocolate and pale lilac in this case) in each generation.

EXPERIMENTAL PRODUCTION OF THE e – g – D LOCI COLOUR COMBINATIONS. Examples of each of the colour combinations just described were produced experimentally in the F_2 generation from a single mating involving a Khaki Campbell drake (eeggdd) and a pale Blue Orpington duck (EEGGD–). These results are reported in Table XV. The former breed carries the three factors in a recessive state and in the latter they are all dominant.

These matings (Nos. 17 and 26) together with backcrosses of the F_1 generation to the recessive Khaki Campbell (Nos. 24 and 25) were carried out in 1957 and 1958 to obtain the complete range of colours in both sexes. Although the full range of colours was obtained, one of the sexes was deficient in a few cases. These have been observed, however, in other matings.

THE BEHAVIOUR OF BLUE AND BROWN DILUTIONS IN A SPECIES CROSS. It is well known that hybrids between different species bearing the "wild-type" pattern are usually intermediate in character. When one of the patterns is modified by the action of a single mendelian factor, however, the effect on the hybrid is similar to the effect on the parent species bearing it. Several instances of this in ducks can be found in the literature: PHILLIPS (1921) showed that when the dusky pattern (m^d) of the Mallard (*Anas platyrhynchos*) was brought into contact with the germ plasm of two other species of duck [The American Black duck (*Anas rubripes*) and the Australian Black duck (*Anas superciliosa*)] it behaved in much the same way as in the pure Mallard. SOKOLOVSKAJA (1935) also demonstrated that brown dilution (d) from the Khaki Campbell male, which was crossed with a Muscovy female (*Cairina moschata*), behaved in the same way as if it were crossed with a Rouen female, namely that the male offspring were dark and the females light brown.

To investigate the behaviour of a factor thoroughly in a species cross, either the hybrids must be fertile to produce an F_2 generation, or, as in the case of most sterile intergeneric crosses, a recessive factor must be sex-linked and a dominant should be easy to recognize phenotypically according to whether it is homozygous or heterozygous. For this reason the male chosen for mating with a Black Muscovy female at the National Institute of Poultry Husbandry in 1958 was a dark Buff Orpington of the genotype: Ggdd.

The results, which are given in Table XIX, can be summarized as follows: All the offspring at day-old were of the same basic pattern that is similar to the "wild pattern" of the Mallard. As expected the progeny were of four different colours. Of the 22 progeny hatched 7 of the males were of the dark unmodified "wild type" colour (ggDd); 4 of the males were modified only by the heterozygous expression of blue dilution (GgDd); 7 of the females were modified by recessive brown dilution in the hemizygous state (ggd–); and the remaining 4 females were modified by heterozygous blue and hemizygous brown dilutions combined (Ggd–) – *Plate 11*. At maturity the "wild-type" males were intermediate in appearance between the two parent species. The presence of brown markings on the hybrids suggests that the Muscovy is homozygous for non-extended black (e). The blue diluted hybrids were intermediate in appearance between "heterozygous" Blue Muscovies and "heterozygous" Blue Mallards. The brown diluted hybrids were analogous in colour to the Khaki Campbell and the blue and brown combinations resembled the Buff Orpington in colour.

These results confirm SOKOLOVSKAJA's findings and also show that blue dilution in the Mallard has a corresponding recessive allele in the Muscovy.

In 1961 and 1962 four further matings (Numbers 74, 75, 84 and 85) were carried out between *A. platyrhynchos* (Blue and Black Orpingtons) and *C. moschata* (Blue and Black Muscovies) to determine whether the blue dilutions of the two species were caused by the same gene and locus. Details of these matings are given in Table XX. The results indicate that the genes concerned are present at different loci in the two species since only the effect of blue dilution of the common duck is evident in the hybrids. It would also appear that because the blue dilution gene of the Muscovy is masked by the genome of the common duck it has a different origin from its counterpart in the Blue Orpington. APPLEYARD (1949) lists amongst the various colours of the Muscovy duck a dun (brown) variety. The mode of inheritance of this colour in the Muscovy does not appear to have been investigated. It would be interesting to see whether it is sex-linked and if so whether it is the same gene as that responsible for brown dilution in *Anas platyrhynchos*.

BUFF DILUTION
(*plate 12*)

Buff dilution is a sex-linked recessive characteristic. Its presence was first suspected in two ways:

(1) Whenever blue and brown dilutions were brought together from independent sources, other than the Buff Orpington, in the presence of e, the resulting buff duck was always found to be darker than the pure Buff Orpington in down, plumage, bill and leg colour. The production of a buff by this means is demonstrated in Table XV.

(2) Recessive segregates (ggd(d)) from Standard Buff Orpington matings (Ggdd × Ggd–), which should theoretically have appeared identical to the Khaki Campbell in down, plumage, bill and leg colour, were always paler in these areas.

Confirmation was provided when Pale Buff Orpington and Khaki Campbell ducks were crossed reciprocally to demonstrate the heterozygous nature of dark buff. The offspring from these matings were expected to be standard (dark) buff in colour and to be uniform in all cases. Instead, matings 41 and 42 (Buff Orpington male × Khaki Campbell female) produced ducklings which were about equally divided into two different shades (Table XXI).

The two shades were still intermediate between the day-old down colours of the "homozygous" parents but this dimorphism of down colour indicated that another factor was involved. On sexing these ducklings all the dark individuals were found to be males and all the pale ones females. The difference was quite decisive and no errors were made in classification. When the reciprocal mating (no. 43) was carried out (Khaki Campbell ♂ × Pale Buff Orpington ♀), all the offspring, male and female, were dark, resembling the males of the previous matings and thus confirming the dominance of the darker shade. These and other matings giving identical results are summarized in Table XXI.

It would appear from the above results that a sex-linked recessive character is present in coupling phase with brown dilution (d) in the

Buff Orpington. The symbol bu is suggested for this factor. In the Khaki Campbell brown dilution is associated with the dominant allele (Bu) in repulsion phase.

The reason why the effects of buff dilution have not been observed before is that in sex-linked matings involving the Bu–bu relationship the Bu–bu factor pair is nearly always present in coupling phase with D and d. The expression of bu, therefore, is masked by and mistaken for the effects of d with which it is linked. This explains why sex-linked crosses involving the Buff Orpington always produce greater colour differences between the sexes than when other brown-diluted varieties are used such as the Khaki Campbell (ddBuBu). Both factors (d and bu) are diluters of black pigment and, therefore, work together additively. In the past there was little to gain commercially from crossing the Buff Orpington with the Khaki Campbell since both varieties were good egg producers and both carried the only sex-linked factor (d) that was then known.

At day-old the dark males (GgddBuBu) can be distinguished from the pale females (Ggd–bu–) by their darker down colour. To confirm this classification one is advised to examine the bill, legs and feet which are always darker in the non-diluted (Bu) birds. Khaki segregates from pure Buff Orpington matings [ggd(d)bu(bu)] can be differentiated from pure Khaki Campbells [ggd(d)Bu(Bu)] in the same way except that the difference in down colour is less marked and more reliance on bill, legs and feet colour is necessary.

At maturity non-diluted (Bu) birds differ from diluted (bu) birds of the same sex in their darker plumage, bill, legs and feet. In birds carrying the heterozygous blue genotype (Gg) Bu allows an appreciable amount of slate colour to be retained in and amongst the buff feathers whereas the recessive dilution factor (bu) completely prevents this slatiness and produces a much paler type of buff. Although the difference is less marked in the non-blue genotype (gg), it is still apparent particularly on the breast where buff dilution produces a paler colour. In some birds the bean on the tip of the bill shows a greater colour difference than the bill itself. It is not possible to compare mature males of one phenotype (Bu) with mature females of the other (bu) since the normal sex-limited differences in colour, (controlled by sex hormones), in birds of the same genotype often outweigh or confuse the differences due to buff dilution (bu) and non-buff dilution (Bu), *e.g.* the male's bill in most coloured varieties is paler and sometimes greener than the female's. The legs and feet in some varieties, particularly the Khaki Campbell and Buff Orpington are darker in the female. These sex differences tend to counteract the genetic differences due to buff dilution because they are working in the opposite direction. Sex dimorphism of plumage pattern is another factor which causes confusion by giving the male a different basic appearance from the female. However, these difficulties do not detract from the value of this factor in sex-linkage since the sexes of coloured varieties can easily be distinguished by voice and plumage pattern at maturity. It is at day-old where sex-linkage is most valuable since it eliminates the need for vent sexing. It has already been shown that the sex-linked Bu–bu differences are 100% reliable for separating the sexes at this age.

No work has been done on the effect of buff dilution on extended black in the absence of brown dilution. The only indication that some modification would occur is found in the lilac females resulting from a Pale Buff Orpington ♂ × Black Orpington ♀ mating. These lilac females, if compared with lilac females resulting from a Khaki Campbell ♂ × Blue-splashed White Orpington ♀ mating are seen to be much paler due to the presence of hemizygous buff dilution.

In 1961 mating No. 76 was carried out to test the linkage strength between the two sex-linked dilution genes. The genotypes of the parents were as follows:

$$gg(D\ Bu)\ (d\ bu)\ \male \times 3\ GG\ (d\ bu)\ (-\ -)\ \female\female$$

Out of a total of 268 ducklings hatched no cross-overs could be detected indicating either very close linkage or allelism. In the former case it should be possible to reveal the missing repulsion phase combination (Dbu). This should prove a new and interesting colour variety since the effect of bu has so far never been observed independently of d. A further variation would be observed if extended black were to be substituted for non-extended black against the background of the new combination. However, the multiple allelomorphism of D, d and bu (in that order of dominance) cannot be entirely ruled out until sufficient numbers of offspring have been hatched from test matings to detect the missing cross-over combination or until more direct proof of very close linkage is found.

The similarity of action of d and bu would perhaps tend to support the existence of a triple series but it remains merely an academic point until further evidence becomes available.

ECONOMIC VALUE AND RANGE OF APPLICATION OF SEX-LINKAGE IN THE DUCK. Sex-linked matings in the duck have little application in the commercial field due to the simplicity of hand sexing. Personnel can be trained to sex ducklings accurately by the vent method within a few hours thus eliminating the need for sex-linked colour matings.

From an academic point of view all varieties of common duck can be used in sex-linked matings provided the right breed is selected as a partner. An extensive list of combinations of varieties for use in sex-linked matings is given in Tables XVI and XVII. All but one of these crosses rely on the D–d factor pair. The exception is the Buff Orpington – Khaki Campbell mating which depends on buff dilution (bu) exclusively. Of all the matings involving brown dilution (d) in Table XVII the only ones which cannot be used for differentiation at day-old are those where both parents are homozygous for blue dilution (G). When one of the parents is homozygous and the other heterozygous for G only half of the offspring can be sexed at sight. As stated, these disadvantages only apply at day-old; in the growing duckling and adult bird the D–d relationship has a universal application. Where heterozygous blue is used in one parent with a non-blue diluted partner the sexes can still be separated at day-old but there are two types of male and two types of female (Table XIX). Although white breeds are normally homozygous for D, it may be necessary to test cross a few birds first to confirm this before carrying it out on a large scale.

To simplify Table XVII all lilac, buff, blue and blue mallard parents are assumed to be homozygous for blue dilution (G). In their present form Tables XVI and XVII give 161 different breed combinations. It is possible that most other breeds carrying brown dilution can replace the Khaki Campbell in the last mating of Table XVII, viz. the Chocolate, Khaki and Lilac group breeds, thus increasing the possible number of combinations still further.

The recommendations regarding sex-linked matings in Tables XVI and XVII are based on experimental findings which are presented in Tables XV, XVIII and XXI. The appearance of these day-old down colour combinations have already been described. Although not all the sex-linked matings included in Table XVII have actually been carried out, the expected results of the missing ones have been observed in the day-old down colours of the offspring from mating 26 (Table XV). The only reciprocal matings to be carried out were between males of the Mallard and White groups and females of the Khaki group (matings 10, 50, and 51 – Table XVIII), and between Khaki Campbell males and Buff Orpington females (matings 43 and 44 – Table XXI).

SUMMARY

(1) The mode of inheritance and phenotypic effects of extended black (E), blue dilution (G), and sex-linked brown dilution (d) on the "wild-type" down and plumage pattern of the common duck are summarized.

(2) Further colour variations, caused by differences at the e, g and D loci in all possible combinations, have been produced experimentally and are described at hatching and at maturity.

(3) Blue and brown dilutions of the common duck (*Anas platyrhynchos*) are shown to have corresponding alleles in the genome of the Muscovy duck (*Cairina moschata*). The behaviour of these dilution factors in a species cross is discussed. Breeding tests indicate that the blue dilution genes of the common duck and the Muscovy have different origins and are situated at different loci on the chromosomes.

(4) Experimental and observational evidence is presented to demonstrate the presence of a further sex-linked recessive gene, buff dilution (bu), in the Buff Orpington duck. The phenotypic effects of this new dilution gene on the following genotypes is described: ee Gg d (d) and ee gg d (d). Linkage tests between d and bu have so far failed to reveal any cross-over progeny.

(5) By means of experimental crosses brown dilution has been shown to have an almost universal application in sex-linked colour matings of the duck. Tables are presented, including all the common breeds of duck, to show the expected results from sex-linked matings involving brown dilution and buff dilution.

TABLE X

SHOWING THE INHERITANCE OF EXTENDED BLACK

Mating Number	Parents			Phenotypes of progeny	
	Breeds	Genotypes		Black (E)	Non-black (e)
16	♂Black Cayuga	EE		9	—
	♀White Campbell	ee			
17	♂Khaki Campbell	ee		58	—
	♀Blue Orpington	EE			
18	♂Buff Orpington	ee		24	—
	♀Black Orpington	EE			
19	♂Buff Orpington	ee		26	—
	♀Blue Orpington	EE			
20 & 21	♂Khaki Campbell	ee		47	—
	♀Black Orpington	EE			
22 & 23	♂Khaki Campbell	ee		18	—
	♀Black Cayuga	EE			
	Total			182	—
	Expected			182	—
24	♂F₁ – mating 17	Ee		42	49
	♀Khaki Campbell	ee			
25	♂Khaki Campbell	ee		22	7
	♀F₁ – mating 17	Ee			
28*	♂F₁ – mating 16	Ee		30	23
	♀White Campbell	ee			
	Total			94	79
	Expected			86.5	86.5
	$\chi^2 = 1.301$	$0,2 < P < 0,3$			
26	♂F₁ – mating 17	Ee		44	17
	♀F₁ – mating 17	Ee			
27*	♂F₁ – mating 16	Ee		100	42
	♀F₁ – mating 16	Ee			
	Total			144	59
	Expected			152.25	50.75
	$\chi^2 = 1.788$	$0,1 < P < 0,2$			

* Only coloured progeny included in figures for these matings.

TABLE XI

THE GENETICS OF BLUE DILUTION IN POULTRY

Species		Dominant or Recessive	Sex-linked or Autosomal	Symbols	Main Authorities
Common Duck	*Anas platyrhynchos*	Partial* Dominant	Autosomal	G(Bl)	Jaap & Milby (1944), Jaap & Hollander (1954).
Muscovy Duck	*Cairina moschata*	Partial* Dominant	Autosomal	G(n)†	Taibel (1954).
Turkey	*Meleagris gallopavo*	Partial* Dominant	Autosomal	G(D)	Robertson, Bohren & Warren (1943), Jaap & Milby (1944), Jaap & Hollander (1954).
Turkey	*Meleagris gallopavo*	Recessive	Autosomal	sl	Asmundson (1940).
Common Fowl	*Gallus domesticus*	Partial* Dominant	Autosomal	Bl	Bateson & Punnett (1906), Lippincott (1918).
Common Fowl	*Gallus domesticus*	Partial* Dominant	Sex-linked	Sd	Munro (1946), Hutt (1949). Van Albada & Kuit (1960).
Common Fowl	*Gallus domesticus*	Recessive	Autosomal	pk (Pink eye)	Warren (1940), Hutt (1949).
Guinea Fowl	*Numida meleagris*	Recessive	Autosomal	i	Ghigi (1924 & 1927).
Common Goose	*Anser anser*	Partial* Dominant	Sex-linked	Sd	Jerome (1953).
Common Pigeon	*Columba livia*	"Wild type" blue**		+	Levi (1957).
Common Pigeon	*Columba livia*	Partial* Dominant	Autosomal	In (Indigo)	Levi (1957).

* In each case the effect of this dilution factor is greater in the homozygous than the heterozygous state. The heterozygotes all show black or dark flecking with the exception of the Muscovy. The work of *Van Albada* and *Kuit* (1960) indicates that the sex-linked dilution gene of the fowl (Sd) may be allelic with sex-linked barring (B).

† TAIBEL suggests n for this factor by ascribing to its allele the property of an incomplete dominant which allows full intensity of black pigment. Since it behaves in exactly the same manner as the majority of the above the symbol G is suggested to avoid inconsistency.

** Recessive to autosomal black (S).
Dominant to autosomal solid red (e).
Recessive to sex-linked ash red (B^a) ⎫
Dominant to sex-linked brown (b)   ⎭ Alleles.

Added in proof: HOLLANDER and WALTHER (1962) describe a further type of blue dilution in the Muscovy. This character is an autosomal recessive and has been allotted the symbol *l* (lavender).

TABLE XII

SHOWING THE INHERITANCE OF BLUE DILUTION IN THE PRESENCE OF EXTENDED BLACK

Parents			Phenotypes of Progeny		
Mating Number	Breeds	Genotypes	Blue-Splashed White	Blue	Black
29, 30 & 31	♂Blue Orpington	GgEE	24	53	25
	♀Blue Orpington	GgEE			
34	♂F₁ – mating 32	GgEE	13	36	12
	♀F₁ – mating 32	GgEE			
	Total		37	89	37
	Expected		40.75	81.5	40.75
	$\chi^2 = 1.380$	$P = 0.5$			
32	♂Black East Indian	ggEE	—	7	15
	♀Blue Orpington	GgEE			
33	♂Blue Orpington	GgEE	—	27	26
	♀Black Orpington	ggEE			
35	♂Blue Orpington	GgEE	—	11	13
	♀Black Cayuga	ggEE			
	Total		—	45	54
	Expected		—	49.5	49.5
	$\chi^2 = 0.818$	$0.3 < P < 0.5$			
36 & 37	♂Splashed-White Orpington	GGEE	—	86	—
	♀Black Orpington	ggEE			
	Expected		—	86	—
38	♂Splashed-White Orpington	GGEE	22	—	—
	♀Splashed-White Orpington	GGEE			
	Expected		22	—	—

TABLE XIII

EVIDENCE OF INDEPENDENT ASSORTMENT BETWEEN BLUE DILUTION (G) AND EXTENDED BLACK (E)

Mating Number	Parents Breeds	Genotypes	GE	Ge	gE	ge
			Phenotypes of Progeny			
17	♂ Khaki Campbell	eegg	58			
	♀ Blue Orpington	EEGG				
		Expected	58			
26	♂ F$_1$ – mating 17	EeGg	32	12	12	5
	♀ F$_1$ – mating 17	EeGg				
		Expected	34.3125	11.4375	11.4375	3.8125
	Linkage – $\chi^2 = 0.046$	$0.8 < P < 0.9$				
24	♂ F$_1$ – mating 17	EeGg	20	22	22	27
	♀ Khaki Campbell	eegg				
25	♂ Khaki Campbell	eegg	10	5	12	2
	♀ F$_1$ – mating 17	EeGg				
		Total	30	27	34	29
		Expected	30	30	30	30
	Linkage – $\chi^2 = 0.033$	$0.8 < P < 0.9$				

TABLE XIV

SHOWING THE INHERITANCE OF BLUE DILUTION IN THE PRESENCE OF NON-EXTENDED BLACK (e) AND BROWN DILUTION (d)

Mating Number	Parents Breeds	Genotypes	Pale Buff	Dark Buff	Khaki
			Phenotypes of Progeny		
39 & 40	♂ Dark Buff Orpington	Ggeedd	18	35	20
	♀ Dark Buff Orpington	Ggeed–			
		Expected	18.25	36.5	18.25
		$0.8 < P < 0.9$			
		$x^2 = 0.234$			
41 & 42	♂ Pale Buff Orpington	GGeedd	—	104	—
	♀ Khaki Campbell	ggeed–			
43	♂ Khaki Campbell	ggeedd	—	21	—
	♀ Pale Buff Orpington	GGeed–			
		Total	—	125	—
		Expected	—	125	—
44	♂ Khaki Campbell	ggeedd	—	22	19
	♀ Dark Buff Orpington	Ggeed–			
		Expected		20.5	20.5
		$0.5 < P < 0.7$			
		$x^2 = 0.220$			
45	♂ Non-diluted Buff Orpington	ggeedd	—	—	24
	♀ Khaki Campbell	ggeed–			
		Expected	—	—	24

TABLE XV

SHOWING VARIETY OF COLOURS OBTAINED FROM MATINGS INVOLVING THREE
FACTORS

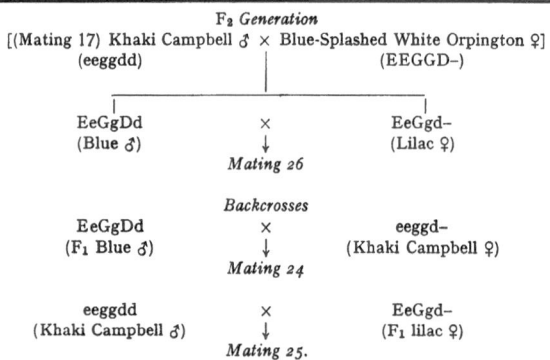

F₂ *Generation*
[(Mating 17) Khaki Campbell ♂ × Blue-Splashed White Orpington ♀]
(eeggdd)   (EEGGD–)

EeGgDd	×	EeGgd–
(Blue ♂)	↓	(Lilac ♀)
	Mating 26	

Backcrosses

EeGgDd	×	eeggd–
(F₁ Blue ♂)	↓	(Khaki Campbell ♀)
	Mating 24	

eeggdd	×	EeGgd–
(Khaki Campbell ♂)	↓	(F₁ lilac ♀)
	Mating 25.	

Phenotype	Genotypes Males	Genotypes Females	Mating 26 Observed	Mating 26 Expected (Approx.)	Mating 24 Observed	Mating 24 Expected	Mating 25 Observed	Mating 25 Expected
Black	{ EEggDd EeggDd	EEggD– EeggD–	9	6	11	11.375	—	—
Blue	{ EEGgDd EeGgDd	EEGgD– EeGgD–	13	12	10	11.375	—	—
Blue-Splashed White	{ EEGGDd EeGGDd	EEGGD– EeGGD–	2	6	—	—	—	—
Chocolate	{ EEggdd Eeggdd	EEggd– Eeggd–	3	6	11	11.375	12	7.25
Lilac	{ EEGgdd EeGgdd	EEGgd– EeGgd–	12	12	10	11.375	10	7.25
Lilac-Splashed White	{ EEGGdd EeGGdd	EEGGd– EeGGd–	5	6	—	—	—	—
Mallard Dark Blue	eeggDd	eeggD–	3	2	15	11.375	—	—
Mallard Pale Blue	eeGgDd	eeGgD–	3	4	13	11.375	—	—
Mallard	eeGGDd	eeGGD–	1	2	—	—	—	—
Khaki	eeggdd	eeggd–	2	2	12	11.375	2	7.25
Dark Buff	eeGgdd	eeGgd–	3	4	9	11.375	5	7.25
Pale Buff	eeGGdd	eeGGd–	5	2	—	—	—	—
Totals			61	64	91	91	29	29

TABLE XVI

CLASSIFICATION OF COMMON VARIETIES OF DUCK FOR SEX-LINKED MATINGS

Black Group [EEggD(D)]
Black Orpington
Black Cayuga
Black East Indian
Black Indian Runner
Black and White Magpie

Mallard Group [eeggD(D)]
Mallard
Rouen
Brown Decoy (Call)
Silver Appleyard
Pencilled Indian Runner

Blue Group [EEGGD(D)]
Blue Orpington
Blue Swedish

Blue Mallard Group [eeGGD(D)]
No recognised variety bearing this
  colour combination.

Chocolate Group [EEggd(d)]
Chocolate Orpington
Chocolate Indian Runner

Khaki Group [eeggd(d)]
Khaki Campbell
Fawn Indian Runner
Fawn and White Indian Runner
Welsh Harlequin

Lilac Group [EEGGd(d)]
No recognized variety bearing this
  colour combination.

Buff Group [eeGGd(d)]
Buff Orpington

White Group [eeggD(D)cc]
Aylesbury
Pekin
Penine
White Campbell
White Orpington
White Indian Runner
White Decoy

TABLE XVII

COMPLETE LIST OF SEX-LINKED MATINGS IN THE DUCK

	Male Parent Group	Female Parent Group	Colour of Progeny	
			Males	Females
D–d Series	Chocolate	Black Mallard White	Black	Chocolate
	Khaki	Black		
	Chocolate	Blue Blue Mallard	Blue (Gg)	Lilac (Gg)
	Lilac	Black Mallard White		
	Khaki	Blue		
	Buff	Black		
	Lilac	Blue Blue Mallard	Blue (GG)	Lilac (GG)
	Buff	Blue		
	Khaki	Mallard White	Mallard	Khaki
	Khaki	Blue Mallard	Blue Mallard (Gg)	Buff (Gg)
	Buff	Mallard White		
	Buff	Blue Mallard	Blue Mallard (GG)	Buff (GG)

	Male Parent	Female Parent	Male Progeny	Female Progeny
Bu–bu Series	Buff Orpington	Khaki Campbell	Buff (GgBubu) (Dark)	Buff (Ggbu–) (Pale)

TABLE XVIII

TO DEMONSTRATE THE VERSATILITY OF THE BROWN DILUTION FACTOR (d) IN
SEX-LINKED MATINGS

Parents			Phenotypes of Progeny			
Mating Number	Breeds	Genotypes	Males	Females	Males	Females
			Blue		*Lilac*	
17	{ ♂Khaki Campbell ♀Blue-Splashed White Orpington	eeggdd EEGGD–	32	—	—	26
18	{ ♂Pale Buff Orpington ♀Black Orpington	eeGGdd EEggD–	15	—	—	9
			47	—	—	35
			Blue-Splashed White		*Lilac-Splashed White*	
19	{ ♂Pale Buff Orpington ♀Blue-Splashed White Orpington	eeGGdd EEGGD–	12	—	—	14
			Black		*Chocolate*	
20 & 21	{ ♂Khaki Campbell ♀Black Orpington	eeggdd EEggD–	24	—	—	23
22 & 23	{ ♂Khaki Campbell ♀Black Cayuga	eeggdd EEggD–	9	—	—	9
			33	—	—	32
			Blue Mallard		*Dark Buff*	
46	{ ♂Pale Buff Orpington ♀Rouen	eeGGdd eeggD–	6	—	—	5
			Mallard		*Khaki*	
47	{ ♂Khaki Campbell ♀White Campbell	eeggdd eeggD–	28	—	—	24
48	{ ♂Welsh Harlequin ♀Rouen	eeggdd eeggD–	24	—	—	18
15	{ ♂Welsh Harlequin ♀White Campbell	eeggdd eeggD–	14	—	—	15
49	{ ♂Khaki Campbell ♀Rouen	eeggdd eeggD–	16	—	—	19
			82	—	—	76
			Mallard		*Khaki*	
50	{ ♂White Campbell ♀Khaki Campbell	eeggDD eeggd–	12	12	—	—
10	{ ♂Rouen ♀Welsh Harlequin	eeggDD eeggd–	25	27	—	—
51	{ ♂Rouen ♀Khaki Campbell	eeggDD eeggd–	8	10	—	—
			45	49	—	—

TABLE XIX

HYBRIDIZATION EXPERIMENT TO SHOW THE EFFECTS OF BROWN AND BLUE
DILUTIONS OF THE COMMON DUCK (*Anas platyrhynchos*) ON THE MUSCOVY DUCK
(*Cairina moschata*)

Mating 52	Dark Buff Orpington (Ggdd) × Black Muscovy (ggD–)		
Genotypes of Progeny	*Phenotypes of Progeny*	*Males*	*Females*
ggDd	Intermediate between wild types of parent species	7	—
GgDd	Blue diluted version of above (≡ Blue Mallard)	4	—
ggd–	Brown diluted version of above (≡ Khaki)	—	7
Ggd–	Brown and blue diluted version of above (≡ Buff)	—	4
	Totals	11	11

TABLE XX

SHOWING THE BEHAVIOUR OF TWO DIFFERENT KINDS OF BLUE DILUTION IN A
SPECIES CROSS

	Parents		Phenotypes of Progeny		
Mating No.	Breeds	Genotypes	GG	Gg	gg
74	♂Pale Blue Muscovy	GG	—	8	9
	♀Dark Blue Orpington	Gg			
75	♂Dark Blue Orpington	Gg	—	14	14
	♀Pale Blue Muscovy	GG			
	Total		—	22	23
84	♂Pale Blue Muscovy	GG	—	—	32
	♀Black Orpington	gg			
85	♂Black Muscovy	gg	—	10	—
	♀Pale Blue Orpington	GG			

TABLE XXI

TO DEMONSTRATE THE SEX-LINKED NATURE OF BUFF DILUTION (bu)

Mating Number	Parents			Phenotypes of Progeny			
	Breeds	Genotypes		Dark Shade		Light Shade	
				Males	Females	Males	Females
41 & 42	♂Pale Buff Orpington ♀Khaki Campbell	GGddbubu ggd–Bu–		63	—	—	41
45	♂Non-diluted Buff Orpington ♀Khaki Campbell	ggddbubu ggd–Bu–		12	—	—	12
	Total			75	—	—	53
44*	♂Khaki Campbell ♀Dark Buff Orpington	ggddBuBu Ggd–bu–		24	17	—	—
43	♂Khaki Campbell ♀Pale Buff Orpington	ggddBuBu GGd–bu–		12	9	—	—
	Total			36	26	—	—

* Mating 44 produced two major shades of colour due to the presence of heterozygous blue dilution Gg and gg phenotypes). Since the dark shade (Bu phenotype) males from matings 41–42 (Gg) and 45 (gg) were available for comparison it was possible to classify these progeny according to their Bu or bu genotype. The effects of blue dilution, therefore, are ignored in the above table.

CHAPTER III

THE INHERITANCE OF WHITE AND WHITE MARKINGS

WHITE MARKINGS IN THE WILD PATTERN

White markings which normally occur in the wild pattern of the Mallard duck are well-defined and restricted to a few specific areas: *Male (Breeding plumage)* – Narrow white neck ring, four white stripes (two on each wing) bordering the wing specula, white ventral wing surfaces and mainly white rectrices.

Female, Juvenile and Eclipse Male – White specular stripes and white ventral wing surfaces only.

The distribution and extent of white areas in the wild pattern are influenced by the four major pattern genes which were referred to in Chapter I of this paper. The presence of the dusky gene (m^d) causes the neck ring and ventral wing surfaces to become pigmented. The restricted (M^R), light phase (li) and harlequin phase (li^h) genes, in addition to the white markings mentioned above, are also responsible for the presence of white in other areas.

The normal white areas of the wild pattern are sometimes slightly modified in the domesticated varieties. For instance, there is less white in the tail of the Rouen than in that of the Mallard. In wild birds, however, these areas are remarkably constant.

All white areas of the wild pattern are obscured by the presence of extended black (E) and recessive white (c). In the presence of extended black, however, other white markings can occur which are not hypostatic to E. These variations are usually quite simply inherited and can also occur in the presence of non-extended black (e). In contrast to the white areas of the unmodified wild Mallard the latter markings often show considerable variation in shape and size. Although the presence of these markings is primarily decided by one major gene, they are also influenced by associated minor modifiers which are susceptible to selection. This aspect is examined in greater detail below.

The following sections review the occurrence and mode of inheritance of all common forms of white and white spotting in the duck.

DOMINANT BIB (S)

MODE OF INHERITANCE. Dominant "bib" does not appear to have been investigated before. It occurs in the following pure breeds: Blue Orpington, Black Orpington, Chocolate Orpington and Blue Swedish. The white markings roughly occupy the area of the breast normally covered by claret brown in the Mallard male and extend upwards on to the ventral surface of the neck (*Plate 13*). Great variation in size and shape is apparent.

To investigate the genetics of this trait a Khaki Campbell (non-bibbed) male was mated with a Blue Orpington (bibbed) female (mating 17). As expected all the F_1 generation birds from this mating possessed well-defined bibs. The F_2 generation (mating 26) gave a close approximation to a 3 : 1 ratio of bibbed to non-bibbed birds. The backcross matings, which were carried out reciprocally to pure Khaki Campbells (matings 24 and 25), produced a 1 : 1 ratio between the two phenotypes. Other matings between bibbed and non-bibbed varieties gave the same results in the F_1 generation as mating 17. These crosses were: Khaki Campbell ♂♂ × Black Orpington ♀♀ (matings 20 and 21) and Buff Orpington ♂ × Black Orpington ♀ (mating 18).

The foregoing results, which are presented in Table XXII, indicate that the presence of this type of bib is controlled by a completely dominant gene. Since the dominant factor was introduced on the side of the female parent it is also autosomal since no evidence of sex-linkage was observed in the offspring. The symbol suggested for dominant bib is S to avoid confusion with the recessive form of bib (b). S was chosen because the bib sometimes vaguely takes the form of an inverted "shield".

AUTOSOMAL LINKAGE. Critical examination of our domesticated breeds of duck regarding the presence of dominant bib brings one to the following conclusions: Although most breeds which are homozygous for extended black (E) are phenotypically bibbed (exceptions are the Black East Indian and Cayuga), no breeds bearing the recessive allele (e) possess the dominant form of bib. A good example of this is found in the Orpington breed where there are several colour varieties. The varieties carrying extended black, namely Chocolate, Blue and Black, are all homozygous for dominant bib (S), whereas the only colour variety homozygous for e, (excluding the White), the Buff Orpington, does not possess a bib.

In the mating described above (No. 17), where a non-bibbed non-extended black (ssee) male was used with a bibbed extended black (SSEE) female, all the F_1 generation ducklings, as expected, were of the extended black, bibbed phenotype. In the F_2 generation and backcross progeny, however, it was noted that all the bibbed birds were extended black in colour (S?E?) and all the non-bibbed ones were non-extended black (ssee). There were no exceptions to this rule in any of these matings (Table XXII). Complete absence of independent assortment under these circumstances can best be attributed to close autosomal linkage.

Unless it can be proved that "dominant bib" is the result of a secondary pleiotropic effect of E, other combinations resulting from crossing-over or independent mutation are theoretically possible. McILHENNY (1941) described several black ducklings which appeared annually amongst a flock of domesticated Mallards (0.45% of 11,972). Since extended black is a dominant characteristic and since all the aberrant blacks were said to have been culled from the breeding flocks each year, they must have been the result of spontaneous mutation (there is no evidence of a recessive form of self-black in the duck). No white markings were recorded in any of the birds that appeared demonstrating that e is apparently capable of mutating independently of s.

Ducklings which hatch from white parents where E is present as a cryptomere, are usually seen to have white bibs when the gene for colour (C) is restored. This indicates their common origin with the ES domesticated varieties. GOODALE (1911) obtained two such females in the F_1 and two females and one male in the F_2 generation from a Rouen-Pekin cross.

The position of the Black East Indian and Cayuga varieties is discussed under the next heading.

THE EFFECT OF MODIFYING GENES. In his classical work on the

"hooded" pattern of rats Castle (MCCURDY, HANSFORD and CASTLE, 1907; CASTLE and PHILLIPS, 1914; CASTLE, 1916, 1919 and 1951) showed that the pigmented hood on a white background was influenced by two different genetic agencies. One of them, which is always present, consists of multiple modifying factors which are unable to produce a pattern themselves but act as modifiers to the action of the recessive gene for "hooded".

"Hooded" in rats is a pattern in which the entire ventral surface posterior to the head is white. Dorsally the pigment is limited to the head and shoulders, comprising the hood, and a mid-dorsal stripe extending from the hood to the tip of the tail. These pigmented areas can be increased or reduced by selection resulting in plus or minus strains depending on the direction of selection of the modifiers. CASTLE found that the modal phenotype in a "plus series" which had undergone sixteen generations of selection had the back completely black and a small amount of the ventral surface also black. The modal type of the "minus series" after 17 generations of selection had a completely white body except for a small amount of pigment on the dorsal surface of the head.

The following are some of the main points brought to light by CASTLE's work:

(1) Selection results in a rapid increase or decrease in the size of the pigmented area. This is accompanied by a reduction in variability. The variability, however, never entirely disappears.

(2) Crosses between modal representatives of plus and minus series resulting from several generations of selection are intermediate in size of hood. From this type of mating there is always more variability in F_2 than F_1.

(3) Crosses with wild rats (non-hooded) tended to increase the pigmented areas of the minus series and decrease these areas in the plus series so that the difference between the two selected strains became less. Repeated backcrossing of the wild rat almost completely eliminated this difference.

It appears, therefore, that in wild rats not bearing a hood these modifiers are not subjected to selection and thus reach a state of moderate equilibrium. A condition of complete stability is never attained, however, since the number of modifiers is too great.

In the duck, the great variability in size and shape of "bib" in a black variety was first described by KAGELMANN (1951), who divided the range of variation into eight grades. The grade showing the least expanse of white had the white restricted to a small patch on the chin and throat. In the highest grade the "bib" extended from the chin and throat under the breast and abdomen to the cloaca.

This variability again indicates the presence of modifiers. Since the presence or absence of bib is determined by one factor S which appears to be affected in its expression by modifying genes its inheritance very closely resembles that of the hood in rats. Without S, therefore, the modifiers are unable to produce a bib regardless of their capabilities in influencing the expression of this gene.

Measurement of Bib Size. Since the bib is very variable in extent and location and often occurs on curved surfaces of the body, its size is very difficult to measure and express numerically. For this reason only samples from each mating were measured, consisting of a limited number of hatches in each case.

The method of measurement was sufficiently accurate to demonstrate relative size and the extreme variability of this character. At 8 weeks of age the outlines of the white markings were traced on to paper and the area thus obtained was measured by a planimeter. Except for mating 31 the figures quoted in the following discussion were all obtained in this manner and are summarized in Table XXIII. The bibs of the ducklings from mating 31 were measured at 24 weeks of age.

The Case of the Black East Indian. Before the full effect of modifiers was appreciated it was assumed that the two aberrant self-black breeds, the Black East Indian and the Cayuga, were either the result of a cross-over producing a repulsion phase combination (Es) or of independent mutation. Although this may be the case in certain strains, two pieces of evidence have shown that in the strains used at this Institute there may be another explanation.

(1) It was noted that about a third of the offspring of pure-bred Black East Indian and Cayuga matings possessed very small white markings in the upper breast region (matings 62 and 63, Table XXII). These areas were very irregular and broken. The size of the bibs of 3 ducklings of this type were measured with the following results: 0.9, 2.8 and 4.7 cm^2.

(2) Some of the birds of these varieties which were completely black in their first year developed white feathers on the breast in their second or subsequent years. In the Cayugas 4 out of 11 which were black in their first year developed small bibs and other white areas in their second season (*Plate 14*). In the Black East Indian the proportion was 3 out of 10.

The presence of these faults has also been recorded by WRIGHT (1902) in the Cayuga and by PHILLIPS (1915) in the Black East Indian.

It was decided to investigate this problem by means of experimental matings:

Non-bibbed Black × *Bibbed Black Mating.* The first series of matings was between a non-bibbed extended black variety and a bibbed extended black variety as follows:

Mating 32 – Black East Indian ♂ × Blue Orpington ♀.

Mating 35 – Blue Orpington ♂ × Cayuga ♀.

As expected the F_1 generations all had white bibs. It was interesting to note, however, that the size of these bibs was much reduced in comparison to the Blue Orpington parents. Although the bibs were very irregular and broken there was not very much variation in size in birds of the same sex. Unfortunately only two typical birds were measured so that the variation cannot be illustrated statistically. In the two specimens examined the bib size was as follows: male – 38.7 cm^2, female – 3.2 cm^2.

When typical specimens of the F_1 generation from mating 32 were mated together (mating 34) the results in the F_2 generation were quite unexpected. Instead of a normal 3 : 1 ratio, 36 birds had bibs but only 3 were without. The variation in bib size and shape was far greater than in any mating yet carried out. In the sample of ten bibs measured the range was from one or two white breast feathers (0.5 cm^2) to one bird where the white area covered most of the head, neck, shoulders, breast and abdomen, and was unmeasurable (*Plate 15*). The largest of the remaining eight bibs measured occupied an area of 134.3 cm^2 (Table XXIII).

Non-bibbed Non-black × *Non-bibbed Black.* The second series of matings consisted of crossing Khaki Campbell males with Cayuga females (matings 22 and 23). Neither breed is phenotypically bibbed but the former is homozygous for e and the latter for E. Again the results were unexpected; 18 ducklings were hatched and all possessed well-developed bibs. The bibs from this pair of matings were much more regular in outline than the previous series. A further difference was that in the first series the white area was interspersed with patches of colour whereas in the second group the white area was solid. All the birds measured were from the same mating (No. 22).

A similar type of mating to the above was carried out by PHILLIPS (1915) when he mated Black East Indians and Mallards reciprocally. He found white bibs present in both the F_1 and F_2 generations but their number and size was not recorded.

It is obvious from these results that the Black East Indians and Cayugas used in the experiments were genetically bibbed but that the bib was prevented from being expressed phenotypically by the presence of a large number of modifiers. In this respect they are equivalent to the minus series in CASTLE's hooded rats. The presence of these modifiers in the Black East Indian and Cayuga varieties was evident in matings 32 and 35 where the bib size of the Orpington was drastically reduced in the F_1 generation followed by great variation in the F_2. In the second series of matings, where bibs appeared in the offspring from matings involving two non-bibbed varieties, the black breed provided the bib factor S whilst the non-black breed supplied the necessary modifiers to allow the bib to be expressed.

Bib size variation in other matings is described below:

Non-bibbed Non-black × *Bibbed Black*. This mating was described earlier to demonstrate the mode of inheritance of the gene responsible for bib. A Khaki Campbell male was crossed with a Blue Orpington female and produced 58 bibbed offspring (mating 17). Their bibs were not measured but it was obvious that they were smaller in size than in the female parent (*Plate 16*). When the offspring from mating 17 were mated inter se (mating 26) the F_2 birds showed a ratio of 3 black bibbed birds to one non-black non-bibbed. Unfortunately it was again not possible to measure their bibs so that no records of size and variability are available. Matings 24 and 25 consisted of backcrosses to Khaki Campbells and yielded equal numbers of bibbed and non-bibbed progeny. 38 bibbed ducklings were measured from mating 24 (F_1 × Khaki Campbell) and 22 from mating 25 (Khaki Campbell × F_1). The measurements are summarized in Table XXIII. Although this mating was carried out reciprocally there was no evidence of sex-linkage in either bib-size or variability. The most striking feature of the figures obtained is their great variability.

Table XXIII shows extreme variability to be a feature of most matings involving this type of white spotting. Even in pure-bred matings involving dominant bib there is no uniformity. This is demonstrated by the measurements taken of the bibs of 11 pure-bred Blue Orpingtons at maturity.

One final mating remains to be described which illustrates the unpredictability of modifiers which are latent due to absence of S or epistasis of recessive white. This was a cross between a Black Cayuga male (EESSCC) and a White Campbell female (eesscc) – mating 16. The latter bird was already known to be homozygous for non-extended black (e). Nine black ducklings were hatched, they all had bibs except one female which was identical to the male parent. A sample of 6 from the first hatch was measured for bib size. In terms of the coefficient of variation variability of bib size was greater than in the progeny of mating 22 (Khaki Campbell × Black Cayuga).

The Influence of Sex. Sex dimorphism of bib size was apparent in most of the matings examined. Table XXIII shows that in several cases mean bib size was at least twice as great in the males as in the females. Although the males weighed up to 8 ounces more than the females at 8 weeks, the difference in body size was insufficient to explain this dimorphism. Sex linkage was not involved since some matings were carried out reciprocally. Where larger samples were examined, (matings 24 and 25), the differences were statistically significant (P = 0.05).

The effect of sex on variability of bib size as measured by the coefficient of variation, which takes relative bib size into account, shows more variability in the females of matings 24 and 25 than the males. The numbers obtained from other matings were too small to yield any useful information.

The Influence of Dilution Factors. The dilution factors G and d were equally represented in both sexes of matings 24 and 25. However, the variation was too great and the numbers of bibbed birds present were too small to give any clear indication whether these dilution factors, present separately or in combination, had any effect on bib size. Only 50% of the offspring from these matings could be measured since the other 50% were genetically non-bibbed.

Similarly, comparisons between the homozygous and heterozygous expressions of S, can not be included owing to lack of data.

DISCUSSION. Although the results of these experiments agree in general with CASTLE's findings in rats, certain differences need to be considered.

Mating 16 (Cayuga × White Campbell) did not produce the same results as matings 22 and 23 (Khaki Campbell × Cayuga) as expected. This anomaly is probably due to the use of two different non-bibbed non-black varieties in the crosses. Since one (or both) of them may have been crossed previously with a bibbed breed selected for bib size they would differ in the number of modifiers they possessed. The duck has been domesticated for such a long time that it is doubtful whether any of the non-extended black breeds remain unchanged from the wild type regarding modifiers of "bib". To duplicate CASTLE's work on this problem, therefore, one would have to use a wild Mallard as the non-bibbed non-extended black variety and take larger samples.

The average bib size from mating 22 (Khaki Campbell × Cayuga) was larger than from the recessive backcross matings – 24 and 25 [Khaki Campbell × (Khaki Campbell × Blue Orpington)]. In other words when the "plus series" was crossed and backcrossed with the "wild type" it produced a smaller type of bib than when the same wild type breed was crossed with the "minus series". This could be due either to the chance effects of picking extreme samples as parents or to the small number of offspring hatched from mating 22.

In spite of these difficulties certain conclusions can be drawn from the present experiments:

(1) The "bib" of the Blue Orpington is inherited very simply; the gene responsible for its presence is dominant and autosomal.

(2) It is very closely linked with extended black (E) in coupling phase.

(3) Dominant bib (S) is greatly influenced by modifying genes whose frequency can be altered, presumably, by selection in either a positive or a negative direction.

(4) In pure breeds positive selection for bib size has resulted in breeds like the Blue Orpington and Blue Swedish whereas negative selection has produced the Cayuga and Black East Indian where the bib is usually phenotypically absent. When breeds from the two

series are crossed bib size is intermediate in the F_1 generation but shows great variation in the F_2.

(5) When non-bibbed non-black and non-bibbed black ducks are mated together they produce an F_1 generation with bibs. Thus the black variety carries the bib factor (S) while the non-black variety supplies the necessary modifiers to allow bib to be expressed.

(6) Bib size exhibits sex dimorphism in that it is usually proportionately larger in the male than in the female.

(7) The above findings do not exclude the possibility of recombination from crossing-over or independent mutation resulting in genotypic non-bibbed black and bibbed non-black varieties.

RECESSIVE BIB (b)

Another form of white "bib" was reported by JAAP (1933a) in some pure Mallard stock. The white area was large and shield-shaped and occupied most of the claret breast region of the male and a similar area in the female. The outline of the bib was different from that of the Blue Swedish duck. It never extended further forward than the neckring area whereas in the Blue Swedish it sometimes extends upwards as far as the chin. A further characteristic of this form of white marking was that its shape and size were remarkably constant (in contrast to S).

As a result of F_1, F_2 and backcross matings JAAP demonstrated conclusively that this type of bib was completely recessive and autosomal and allotted it the symbol b.

A further feature distinguishing b from S is that in this instance b was found in the presence of non-extended black (e). At this Institute the Se combination has never been observed. Since JAAP did not publish the results of any linkage tests it is not possible to say whether b is inherited independently of the e locus.

WHITE PRIMARIES (w)

One or more white primary wing feathers in ducks can result from the presence of one or both of the following factors: "white primaries" (w) and "runner" (R). Since the latter also produces white in other areas of the body it will be examined under a separate heading.

The white primary factor (w) was first described and analysed by JAAP (1933a) who found that it behaved as an autosomal recessive. To be able to observe its full effect it must occur in the presence of "non-runner" (rr), as the runner factor, whether homozygous or heterozygous, is epistatic to it (Table XXIV). In JAAP's experiments "white primary" ducklings were obtained from a strain of pure Mallards. Since white primaries are also found in the Blue Swedish duck as a breed characteristic, it is obvious that w can occur in the presence of either E or e.

In day-old ducklings it is possible to estimate the number of white primaries the adult will have by the amount of white down which occurs at the tip of the wing and the extent of pigmentation of the feet. Ducklings which will eventually have one or two white primaries on each side show triangular-shaped white areas on the webs of the feet and the ends of the toes may also be white. Where the number of white primaries is likely to be four or more on each side the whole of the feet are white but the shanks remain pigmented. Sometimes the bill of the day-old also has yellowish-white areas on the upper surface. Thus, at day-old all the extremities; bill, feet and wing-tips may have white markings. At maturity, however, the white disappears from the feet and bill and only remains on the wings.

THE RUNNER FACTOR (R)

The "runner" pattern was first investigated by JAAP (1933a). The word "runner" was chosen because the pattern, which is about to be described (*Plate 19*), is borne by the Fawn and White Indian Runner variety as a breed characteristic. In the standard pattern of the Fawn and White Runner the head is adorned with pigmented cap and cheek markings. The cap is separated from the cheek markings by an extension of the neck white. The bill is divided from the head markings by a further narrow prolongation of the neck white, $1/8''$ to $1/4''$ wide, extending from the white underneath the chin. The neck is pure white down to where the sex-dimorphic dark neck area in the male ends, *i.e.* to the white neck ring in the male Mallard. From the middle of the breast, that is at a point half way between the front of the breast bone and the legs, another area of white begins extending downwards on to the thighs. The breast white also passes between the legs to beyond the vent. A third area of white is present on the wings covering the primaries, secondaries and lower part of the wing bow. Thus, in the "runner" pattern white is present in three major areas: head and neck; breast, thighs and abdomen; and wings.

The markings, described above, are those which would be expected in an exhibition bird. In commercial stock, however, the pattern shows much variation in the extent and size of the white areas due to modifying genes.

As a result of his breeding experiments JAAP (*loc. cit.*) found that the runner pattern was due to an incompletely dominant autosomal gene R. In these experiments the runner pattern was extracted from the Mallard and Pekin varieties. In the former it was present in the heterozygous condition and in the latter it was hypostatic to recessive white (c). JAAP showed that in the absence of the gene w for white primaries the runner factor R, in the heterozygous state, was also capable of producing white primaries (Table XXIV).

At day-old "runners" show much white in the posterior ventral regions. At this stage it is sometimes difficult to distinguish between "light phase" (li) and "runner" ducklings. Unlike "runners", light phase ducklings do not normally have any white in the wing tips or dorsal neck surface and there is no sharp line of demarcation between white and pigmented areas across the middle of the breast. In the early stages the neck-white is sometimes obscured because the neck region of the young duckling anterior to the neck-ring is normally light in colour. However, it is not always obscured as white is sometimes apparent on the dorsal part of the neck and so can be used in combination with other white markings to identify R.

The bill, legs and feet of the heterozygous runner (Rr) at day-old, show the same amount of depigmentation as that found in the white primaries class (ww) q.v. In homozygous "runners" (RR) the bill, legs and feet have very little pigment at all compared with the particoloured bills and feet of the "white primaries" groups (Rr or ww).

EFFECT OF MODIFIERS. Reference was made earlier to the effect of modifying genes on the runner pattern. Observations at the National Institute of Poultry Husbandry and elsewhere have shown that these modifiers are capable of affecting the phenotypic expression of R in its heterozygous as well as its homozygous form.

In the F_1 generation of a Pekin-Rouen cross GOODALE (1911) found some birds with white primaries and some with a non-sexual white neck-ring. In the F_2 generation, as well as birds with white primaries and white neck rings there were also a few bearing the runner pattern. These were identified by the description and accompanying photograph.

At this Centre similar results have been obtained and considerable variation in the heterozygous expression of R has been observed. Khaki Campbells and White Campbells were crossed reciprocally; mating 47 (Khaki Campbell × White Campbell) and mating 50 (White Campbell × Khaki Campbell). Except for variations at the M locus the F_1 generation was remarkably uniform in colour at eight weeks of age. There were one or two birds, however, with white primary feathers or a white non-sexual neck-ring: mating 47 produced 51 ducklings of which one male only had white primaries; mating 50 produced 24 ducklings of which two males had the white neck ring of non-sexual origin.

At day-old the ducklings from the F_1 generation were wing-banded and the amount of white on their bills, legs and feet was recorded:

Mating 47 – 20 ducklings had solid-coloured feet and bills, 9 had white in the bill only, 1 had white in the feet only, and 22 had white in both bill and feet.

Mating 50 – 23 ducklings had white in both bill and feet, 1 had white in feet only.

It is possible that the presence of white in these areas may be a means of identifying the R factor in the heterozygous form even when modifiers prevent the appearance of white primaries (see below). This hypothesis is based on the findings of JAAP (1933a) who always found white in the extremities of the body when either R or ww were present. Judgement must be reserved on this point, however, until more evidence is available since a further variable was introduced into these matings in the form of heterozygous colour (Cc). Evidence of white primaries (in the form of white wingtips) was not recorded at day-old, but since only one duckling developed white primaries at eight weeks it is unlikely that this character would have been seen in any but this one bird.

Six females were selected from the F_1 progeny of mating 47 and one male from the F_1 generation of mating 50. These birds were originally selected for their M^R, M and m^d phenotypes but on checking back to the day-old records it was found that the male had white patches on the bill and feet, one female had white patches on the bill only and one on the bill and feet, whereas the four remaining females had solid-coloured bills and feet.

At maturity none of the seven birds had either white primaries or a non-sexual white neck stripe and, of course, the non-pigmented areas of the bill, legs and feet had disappeared. Thus, if R was present in any of these selected birds it had become so modified in the adult as to behave as a recessive to non-runner. Proof of the presence of R was found in the F_2 generation when 14 of the 128 coloured offspring were homozygous for R and displayed the full runner pattern. Backcrosses of the 6 females described, to a white male carrying the runner factor produced 33 offspring of which 4 were pure runners.

On crossing the F_2 runners together they were found to breed true (Table XXV). It was noticed that all the pure runners at day-old had non-pigmented feet and bills. When the backcross mating was carried out it produced variable results. A runner female was first backcrossed to a Rouen male producing 15 Rr offspring. Seven of these showed no signs of unusual white markings whatsoever, 5 had a white neck-stripe only (non-sexual) and 3 had a white neck-stripe and white primaries (Table XXV). When a runner male (RRee) was crossed with a Cayuga female (rrEE), 12 offspring were produced; 6 with white primaries and 6 without. Since all the offspring from this latter mating had a dominant white bib (S), inherited from the Cayuga, it was not possible to see whether the neck stripe was present or not.

JAAP (1933a) did not observe the presence of any white in the neck region of heterozygous runners. Neither did he find any heterozygous runners without white primaries. These differences are undoubtedly due to the action of modifying genes.

DISCUSSION. In certain circumstances the runner factor R behaves as an incomplete dominant and exhibits an additive effect according to whether it is homozygous or heterozygous. When heterozygous its effect can be seen as white primaries or a non-sexual white neck ring (or both), present in both sexes. When homozygous it produces the full runner pattern with white in the head and neck; breast, abdomen and thighs; and in the wings.

Sometimes modifiers of the type associated with dominant bib (S) can prevent the penetrance of R in the heterozygous state making it appear as a recessive. In the latter case it is possible that the presence of R may still be detected in the day-old heterozygote by the appearance of white areas in the bill and feet.

The differences between the heterozygous and homozygous expressions of R are greater than any differences due to the action of modifying genes. This is in complete contrast to dominant bib (S) where the wide variation in bib size resulting from the action of modifying genes far outweighs any differences (if they exist) between the homozygous and heterozygous expressions of S.

OTHER FORMS OF WHITE SPOTTING

Many reports of other types of white spotting can be found in the literature. On examination they are usually found to be the result of: modifications of the types already described, combinations of two or more of these types, species hybrids or various forms of mosaicism. Several of these reports as well as possible causes of other irregular types of white spotting are examined below.

CONFUSION BETWEEN SEXUAL AND NON-SEXUAL NECK RING. WALTHER, HAUSCHILDT and PRÜFER (1932) described an apparent association between white neck ring (occurring in both sexes) and brown dilution (d) as follows:

	With White Neck Ring	Without White Neck Ring
Non-diluted	57%	43%
Brown diluted	10%	90%

No differentiation was made between the sex-limited white neck ring of the male and the non-sexual neck ring of both sexes (probably due to Rr). The figures were taken from offspring arising from breeding experiments designed to demonstrate the sex-linkage of d. These experiments resulted in a greater proportion of males amongst the non-diluted birds than amongst the diluted ones:

	Males	Females
Non-diluted	77	29
Brown diluted	56	112

Therefore, if no distinction is made between the two types of neck ring and sex is not taken into account these figures would tend to show an apparent correlation between presence of neck ring and non-dilution.

A further factor which may influence the figures is that the dusky pattern (m^d), which is carried by the Khaki Campbell (one of the parent breeds), obscures the sexual white neck ring of the male. Chance effects of homozygous m^d in the F_2 and later generations could again lead to wrong conclusions.

THE MAGPIE DUCK (*Plate 18*) – The Magpie is a British breed of duck which in recent years has become very rare. The basic pattern consists of a patch of colour or cap on the top of the head and an area of the same solid colour on the bird's back. On the back the colour extends posteriorly, from a line connecting the points where the wing fronts join the body, to a line running just below the tail. It also runs slightly over on to the edge of the wings. The rest of the body including the primary and secondary wing feathers, thighs, rump, abdomen, breast, neck and face is pure white. The basic colour is usually black but can be diluted to blue or dun (chocolate) by the action of G or d.

Unfortunately all attempts to obtain specimens for genetical analysis have failed, even in the last strongholds of Wales where it retained its popularity until quite recently.

One can only hazard a guess as to its origin from photographs. Two possible explanations are suggested to account for this pattern:

1. The pattern of the Magpie duck is very similar to that of the Black and White Muscovy (*Plates 17 and 18*). Thus, if a fertile Muscovy × Common Duck hybrid male occurred at some time, it could be the means of transferring the black and white muscovy pattern to the common duck.

2. The second explanation involves a combination of the pattern genes R and S. Differences between the runner pattern and that of the Magpie can be accounted for by the addition of a slightly modified (plus series) dominant bib.

The latter is the more feasible of the two hypotheses since fertile Muscovy hybrids are extremely rare. Furthermore, the pattern of the Black and White Muscovy is heterozygous resulting from a mating between a coloured bird and a white one (TAIBEL, 1954). It is, therefore, less likely that a pure breed would be founded on a heterozygous pattern even though it has occurred in the Blue and Buff Orpingtons.

During the present season (1962) several ducklings have been produced by the second method. Although these are still in the down and juvenile stages they bear a remarkable resemblance to the Magpie breed.

"RUNNER" MARKINGS. The white markings described by KAGEL-MANN (1951) in a strain of Rouens were probably due to the homozygous and heterozygous expressions of R.

An adult female Mallard described by SAGE (1954a) on the Wilstone reservoir and the ones described by HARRISON (1959) also appear to carry the runner pattern.

SPECIES HYBRIDS. The Mallards with white spotting described by SAGE (1953, 1954b and 1955) are considered to be the result of interspecific hybridization. Sage suggests that the Mallard may have hybridized with the Pintail, Gadwall and Shoveler respectively. The parent species of hybrids can often be identified by the shape and location of the white markings present, for example, the characteristic vertical white neck stripe of the Pintail drake and the white breast of the Shoveler drake.

MOSAICS AND VARIEGATED CONDITIONS. White feathers or patches in an otherwise pigmented bird in the absence of genes for white spotting can have an environmental or genetic origin. HUTT (1949) states that white feathers can sometimes develop in previously coloured fowls as the result of injury to the skin. They can also be induced by excessive doses of thyroxine or can occur spontaneously. In the latter case Hutt found the condition sometimes to be associated with caponisation or an ovarian tumour.

In a discussion on variegated conditions in domesticated birds HOLLANDER (1944) considered patchy mosaics and irregular variegation or flecking to have similar origins. He mentioned that two possible causes were somatic mutation and somatic segregation. He also pointed out that with variegation there is usually an inherited predisposition or instability, which can be attributed to Mendelian factors or to chromosome anomalies.

It is suggested that a common cause of irregular white flecking in coloured ducks is a localised somatic mutation or segregation of the gene for colour (C), which is present in heterozygous state.

However, it is difficult to attribute white markings in the homozygote to the mutation of this gene unless it is accompanied by segregation. In this case either somatic mutation of dominant genes for white spotting (S and R) or a purely environmental or physiological type of depigmentation would seem to be indicated. What appears to be the reverse mutation of C is often seen in white ducks where odd feathers assume the colour carried by the bird as a cryptomere. This can sometimes be used to identify the hypostatic colour or colours without a progeny test.

In heterozygous birds CREW and MUNRO (1938) suggested that two forms of somatic segregation (non-disjunction and the loss of a chromosome) could account for most forms of mosaicism in birds. If these phenomena are present in ducks they must usually occur late in development since "half-and-half" sectorial mosaics (with or without size differences) and gynandromorphs are very rare.

Although certain pied patterns (dominant bib, recessive bib, white primaries and runner) are known to be under genetic control, the possibility of genetic mosaicism, timed to originate at a certain stage in development, cannot be ignored. However, since melanophores are known to migrate from the neutral crest, HOLLANDER (*loc. cit.*) tentatively suggests that a more likely explanation of these pied conditions is that there may be some genetic hindrance to migration of melanophores in certain regions of the embryo.

THE INHERITANCE OF SELF-WHITE PLUMAGE IN THE COMMON DUCK

RECESSIVE WHITE (c). In all the common white breeds of duck self-white plumage is due to an autosomal recessive gene (c), (JAAP, 1933a; MAZING, 1933; JAAP and HOLLANDER, 1954). This has been confirmed at the National Institute of Poultry Husbandry where several coloured breeds were mated with White Campbells. The results of these matings are shown in Table XXVI. The dominant allele (C) allows full expression of any of the pattern, colour or dilution genes which are present. When c is homozygous it is epistatic to all forms of colour.

DOMINANT WHITE. In the common fowl there are two types of dominant white; one autosomal and the other sex-linked. When homozygous, the autosomal form (I) completely inhibits black pigment but not red and buff. The homozygous sex-linked form (Silver – S) inhibits red and buff but not black. No such forms have been reported in the common duck. The nearest approach to a dominant white in the species is the colour of the homozygous blue female (EEGG). However, it can hardly be classed as a dominant white variety since the male is light grey in colour. The white of the homozygous blue female duck is analogous to that of the white breeds of geese such as the Embden and Roman which carry the sex-linked dilution factor, Sd (JEROME, 1953).

Another type of self-white, which occurs in the Muscovy duck and the guinea fowl, is an incompletely dominant form which results in a coloured-and-white mosaic pattern in the heterozygote (TAIBEL, 1954; GHIGI, 1927). This phenomenon has not been observed in Mallard ducks. The possibility of transferring this factor to the Mallard via species hybrids has already been discussed in connection with the Magpie pattern.

COMPLEMENTARY WHITE. A case where two types of recessive white appeared to be complementary was reported by MCILHENNY (1941). White ducklings appeared sporadically in a flock of wild Mallards maintained as a breeding flock for producing game birds. Six of these white birds, (2 males and 4 females), were reared to maturity and mated together the following Spring. More than 60 ducklings, all unexpectedly bearing the wild Mallard pattern, were hatched from this mating. Since white is normally a recessive in ducks there appear to be only two possible explanations for these results:

(1) Both males were sterile and the females mated with full-winged wild Mallard drakes which flew unobserved into the breeding pen.

(2) Two forms of white, complementary to one another, were present in the breeding stock – each supplying a dominant gene required by the other to produce colour.

Both explanations depend very largely on chance since two males were involved. In the first case they would both have to be completely sterile and in the second both would have to carry the opposite type of white to the females. Since only two white males were involved it is possible that the recessive white carried by the females was sex-linked and that carried by the males was autosomal. Unfortunately the work was not carried any further to test these suppositions.

An attempt was made to reproduce these results at the National

Institute of Poultry Husbandry by crossing several white breeds in different combinations. Some of the breeds chosen would not normally be crossed in commercial practice and have probably very rarely been crossed since the breeds originated. The following combinations were carried out to determine whether one or more factors were responsible for recessive white plumage:

Aylesbury ♂ × White Campbell ♀
White Decoy ♂ × White Indian Runner ♀
White Indian Runner ♂ × White Decoy ♀
White Decoy ♂ × White Campbell ♀

The Pekin was not included because the Aylesbury × Pekin mating is often carried out in commercial practice and is known to produce white ducklings.

Table XXVII summarizes the results of these matings, where all the progeny were white. Although this does not disprove the existence of other forms of recessive white it does show that the common white breeds of duck in Great Britain are all of the same type. Complete failure to reproduce MCILHENNY's results is reminiscent of the failure of other workers to repeat the findings of BATESON and PUN-NETT (1906) with fowls. If the whites used in these experiments were genuine recessive whites the missing types must be caused by genes with very low mutation frequencies.

Other gene combinations can occasionally produce pure white or nearly white phenotypes. These are: homozygous blue (EEGG) in the female; an extremely divergent form, female, of the harlequin phase (eeli^{h}li^h); and albinism. In this particular case true albinism can be excluded since the ducklings had pigmented eyes. Partial albinism of the type carried as a sex-linked recessive by both fowls and turkeys may possibly have been mistaken for recessive white since there is usually some pigment present in the eye. However, since this is usually accompanied by "ghost" markings in the plumage it is extremely unlikely. Homozygous blue and harlequin phase females can also be disregarded since the specimens described showed no traces of melanic pigment in the bill or legs. These areas were completely yellow, typical of a recessive white breed of duck.

ALBINISM. True albinism is undoubtedly very rare in Mallard ducks. According to SCOTT (1957) pure albinism has not been recorded in wildfowl. In the lists of recent British literature on European wildfowl (1945–57) by MATTHEWS (1959), and (1957–60) by OGILVIE (1961), there are no references to *true* albinism in Mallards. There is a painting of an albino Mallard female by T. M. SHORTT in KORTRIGHT's "The Ducks, Geese and Swans of North America" (1943), but no reference to the source of information is given.

There are autosomal and sex-linked forms of recessive albinism in both fowls and turkeys. These four factors differ in the amount of melanic pigment they allow to occur in the bird. In some forms vision only is affected whereas in others general viability suffers as well. In the sex-linked type of fowl albinism (al) the gene has no ill effects at all (MUELLER and HUTT, 1941), whereas in sex-linked turkey albinism (n^a) the birds are practically blind and embryonic and juvenile mortality is extremely high (HUTT and MUELLER, 1942). If an even more severe lethal form were found to occur in ducks with up to one hundred per cent embryonic mortality it would explain why albinism has not been reported in this species. Alternatively, the albino locus or loci may simply be less labile in the duck than in other birds.

VARIATION IN DUCKLING COLOUR. There is considerable variation in the down colour of day-old white-breed ducklings. Although the colour is generally described as yellow, a range of variation exists from a rich orange-red (almost as dark as the Rhode Island Red chick) through pale primrose-yellow to creamy-white. These extreme differences are usually seen in comparisons between different breeds and strains, but similar differences, almost as great, can occasionally be found between individuals of the same strain.

In turkeys this variation in colour of day-old whites is due to hypostatic colour patterns (ASMUNDSON, 1945). White turkeys carrying the bronze pattern (b) show extensive areas of buff pigment in the day-old down corresponding to the brown areas in the bronze pattern. The two mutant alleles of b, black (B) and black-winged bronze (b^1) produce a poult which completely lacks this buff pigment and is practically pure white. This difference in appearance of whites at day-old was at one time suggested as a means of distinguishing between the different breeds of white turkey (MILBY, 1943).

LAMOREUX and HUTT (1942) found variations in White Leghorn chicks from pure white through light cream to golden brown. The brown was present as markings on the back of the neck, across the shoulders and along both sides of the breast. Although the inheritance of these differences was not identified with any known hypostatic patterns as in the turkey, it was considered by the two authors concerned that relatively few genes were responsible because sharp differences were established very quickly without deliberate selection and were easily maintained.

Since the main differences in colour in the turkey were due to variations between the extended black and non-extended black genotypes it was postulated that a similar situation could occur in the duck. In this case the wild pattern of the Mallard (e) would produce a darker and redder type of down than extended black (E). To test this hypothesis matings were carried out between White Campbell and Cayuga ducks. In the F_2 and backcross progeny it was anticipated that it would be possible to differentiate between e and E in the day-old recessive white. However, although considerable variation was found to exist it could not be correlated with the expected effects of E and e.

In the adult white turkey the only difference between hypostatic bronze and hypostatic black is in the colour of the iris. The ccbb genotype (carrying bronze) has a brown iris whereas in the ccBB genotype (carrying black) the iris is blue. Ducklings bearing extremes of down colour were reared and their eyes examined at maturity. Also all white ducklings from a mating designed to produce individuals carrying the two hypostatic patterns e and E. No differences in eye colour could be detected.

The main conclusions to be drawn from this experiment are as follows:

(1) As in the case of the White Leghorn fowl the variation in down colour of white breed ducklings appears to be a polygenic effect. That it is probably due to relatively few genes is again demonstrated by the extreme differences between strains which have not been selected for down colour.

(2) If e and E do have any effect on down colour their expression is influenced to such an extent by modifiers that overlapping occurs between their phenotypic expressions and possibly with the effects of other pattern or colour genes. Thus, it is impossible to separate their effects at the phenotypic level.

SNOW-WHITE DUCKLINGS. HUTT (1951) investigated a mutation in chicks which causes the day-old down colour to appear pure white as opposed to the cream or yellow of normal chicks. He found that the condition was due to a simple autosomal recessive gene which he called "snow-white" (sw). At maturity it was not possible to distinguish between "snow-white" and normal birds and at no time did the gene affect skin or eye colour.

A similar condition was observed in Pekin ducklings by KRIZENECKY (1961) where the down colour of the affected birds was snow-white instead of yellow. There were no differences between snow-white and yellow-downed ducklings in the yellow pigmentation of the shanks and bill, the blue pigmentation of the eyes or in the adult birds. Although preliminary trials were carried out on the inheritance of "snow-white" the change in circumstances of the farm prevented their completion. In these preliminary trials "snow-whites" were found to segregate from yellow-downed parents and when mated inter se produced only ducklings with snow-white down. KRIZENECKY concluded that although the condition was inherited as a recessive it was not possible to decide whether or not it was autosomal or unifactorial due to the limited data available.

The only other species of poultry where this condition has been noted is in the White guinea fowl where snow-white down appears to be a species characteristic. No yellow-downed chicks have ever been observed at this Institute. GHIGI (1936) confirms this and states that these guinea fowl chicks look like white snowballs.

Up to the present time no snow-white ducklings have been available in this country for investigation.

THE EFFECT OF THE RUNNER PATTERN IN WHITE BREEDS. Although no definite correlation has been found between general shade of down colour and hypostatic genes in white-breed ducklings an association has been established between a pattern which is completely hypostatic in the adult and a pattern found in the day-old down. In an earlier section of this paper the manifestation and mode of inheritance of the "runner" pattern was described. White areas were found to occur on the head, neck, breast, abdomen, thighs and wings while the rest of the body was pigmented. In coloured birds of this genotype similar white areas can be found in the day-old down. White-breed ducklings bearing the "runner" pattern show corres-

ponding white areas which contrast with the surrounding yellow
pigment. In other words the coloured areas are replaced by yellow
but the white parts remain unaltered. Thus, the runner pattern is
conspicuous at day-old but absent at maturity in the white bird.

The pattern can be seen equally well whether the ducklings are
homozygous or heterozygous for R. This was demonstrated by mating
together two birds homozygous for R but heterozygous for c (No. 64)
to produce whites homozygous for "runner". To produce the heterozygous type a bird homozygous for non-runner but heterozygous
for colour (rrCc) was mated with a pure runner also heterozygous
for colour (RRCc) – No. 65 (Table XXV). At maturity the "runner
whites" were indistinguishable from normal whites.

APPARENT LINKAGE BETWEEN C AND E. Mating 27 has been mentioned before in connection with the inheritance of extended black
and recessive white. The birds comprising this mating consisted of
the progeny from a Cayuga × White Campbell mating (No. 16).
Eight hatches were taken from mating 27 (F_2 generation from
mating 16). Instead of the expected 9 : 3 : 4 ratio between extended
black, non-extended black and recessive white phenotypes, 46 black,
8 non-black and 22 white ducklings hatched. The coloured : white
ratio provided a reasonable fit to the expected 3 : 1 ($\chi^2 = 0.632$;
$0.3 < P < 0.5$), but the extended black : non-extended black ratio
was quite unexpected. The obvious explanation was that linkage
was present between E and C and that the small number of coloured
non-extended black ducklings was the result of crossing-over.

About this time it was realized that errors were being made in
classification due to the presence of the dusky factor (m^d), whose
effect was being mistaken for that of E. This upset the ratios not
only in the dead-in-shell but also in the day-old classifications. It
was, therefore, decided to discard these figures except for the coloured
to white ratios. As soon as the differences between E and m^d were
appreciated in the day-old, further hatches were recorded and all
dead-in-shell excluded.

These later results are presented in Table XXVIII and serve to
demonstrate complete independent assortment between E and C.
Only the coloured progeny were used to detect cross-overs since
it was impossible to identify them in the white birds.

SUMMARY

(1) The distribution of white markings in the wild pattern of the common duck is described. Modifications to the wild pattern by genes occurring at the M and Li loci cause changes in the wild type white spotting.

(2) Investigations into the inheritance of a white breast and neck marking known as "dominant bib" have shown that it is caused by a dominant autosomal gene (S). It is closely linked with extended black (E) and its size is greatly influenced by modifying genes.

(3) Another form of white breast marking called "recessive bib" (b) is compared with the dominant type. It is autosomal and recessive and less influenced by modifiers.

(4) White primary wing feathers are the result of another autosomal recessive gene (w). They can only be observed in the CCrr genotype since the "runner" factor (R) sometimes causes white to occur in the same area.

(5) The runner factor (R) is an incomplete dominant for white spotting found in the Fawn and White Indian Runner variety. It is affected by modifying genes in both the homozygous and hetero-zygous condition.

(6) Other forms of white spotting are discussed. Two hypotheses are put forward to explain the Magpie pattern and several mani-festations of the runner pattern are listed. White spotting resulting from species hybridization and mosaicism is reviewed.

(7) The inheritance of self-white plumage in the duck is summarized. The different types of white discussed are: recessive white, dominant white, complementary white and albinism.

(8) The variation in down colour of white breed ducklings at day- old was investigated. No associations were found between the E, e factor pair and colour of down. Similar situations are described in the turkey and fowl.

(9) The effect of the runner gene on the day-old down colour of white breed ducklings is described.

(10) An apparent case of autosomal linkage between E (extended black) and C (Colour) was investigated. They were found to be inherited quite independently. Previous errors were made through wrong classification of "dusky" phenotypes.

TABLE XXII

RESULTS OF MATINGS DESIGNED TO DEMONSTRATE THE DOMINANCE OF "bib" (S)
AND ITS LINKAGE WITH EXTENDED BLACK (E)

	Parents			Phenotypes of Progeny			
Mating Number	Breeds	Phenotypes	Geno-types	Bib-Black	Bib-Non-black	Non-bib Non-black	Non-bib Black
17	♂ Khaki Campbell ♀ Blue Orpington	Non-bib, Non-black Bib, Black	eess EESS	58			
20 & 21	♂ Khaki Campbell ♀ Black Orpington	Non-bib, Non-black Bib, Black	eess EESS	47			
18	♂ Buff Orpington ♀ Black Orpington	Non-bib, Non-black Bib, Black	eess EESS	24			
			Total Expected	129 129			
26	♂ F₁ – mating 17 ♀ F₁ – mating 17	Bib, Black Bib, Black	EeSs EeSs	44		17	
	$\chi^2 = 0.268$		0.5 < P < 0.7 Expected	45.75		15.25	
24	♂ F₁ – mating 17 ♀ Khaki Campbell	Bib, Black Non-bib, Non-black	EeSs eess	38		47	
25	♂ Khaki Campbell ♀ F₁ – mating 17	Non-bib, Non-black Bib, Black	eess EeSs	22		7	
			Total	60		54	
	$\chi^2 = 0.316$		0.5 < P < 0.7 Expected	57		57	
32	♂ Black East Indian ♀ Blue Orpington	Non-bib, Black Bib, Black	EE?? EESS	22			
35	♂ Blue Orpington ♀ Cayuga	Bib, Black Non-bib, Black	EESS EE??	22			
			Total Expected	44 44			
22 & 23	♂ Khaki Campbell ♀ Cayuga	Non-bib, Non-black Non-bib, Black	eess EE??	18			
			Expected	18			
62	♂ Black East Indian ♀ Black East Indian	Non-bib, Black Non-bib, Black	EE?? EE??	3			6
63	♂ Cayuga ♀ Cayuga	Non-bib, Black Non-bib, Black	EE?? EE??	7			13
			Total	10			19
34	♂ F₁ – mating 32 ♀ F₁ – mating 32	Bib, Black Bib, Black	EE?? EE??	36			3

TABLE XXIII

DEMONSTRATING THE EXTREME VARIABILITY OF DOMINANT BIB SIZE

Parents			Progeny			
Mating Number	Breeds	Sex and Number	Mean (cm²)	Range (cm²)	Standard Deviation (cm²)	Coefficient of Variation %
24	♂Khaki Campbell × Blue Orpington	♂♂(20)	97.5	22.7 to 187.8	46.0	47.2
	♀Khaki Campbell	♀♀(18)	42.0	10.7 to 95.3	27.8	66.1
25	♂Khaki Campbell ♀Khaki Campbell ×	♂♂(11)	97.8	46.2 to 133.2	31.0	31.7
	Blue Orpington	♀♀(11)	42.7	5.9 to 95.5	28.4	66.3
22	♂Khaki Campbell	♂♂(5)	115.1	66.1 to 189.7	46.6	40.5
	♀Cayuga	♀♀(8)	68.7	50.0 to 100.6	19.7	28.8
16	♂Cayuga	♂♂(2)	85.0	26.9 to 143.1	—	—
	♀White Campbell	♀♀(4)	8.6	Nil to 21.1	—	—
31	♂Blue Orpington	♂♂(4)	262.3	94.0 to 400.0	—	—
	♀Blue Orpington	♀♀(7)	211.8	90.2 to 538.9	151.1	71.4
35	♂Blue Orpington	♂(1)	38.7	—	—	—
	♀Black Cayuga	♀(1)	3.2	—	—	—
34	♂Black East Indian × Blue Orpington	♂♂(4)	>86.8	8.0 to >134.3	—	—
	♀Black East Indian × Blue Orpington	♀♀(7)	8.4	Nil to 20.3	7.8	93.0
62	♂Black East Indian	♂(1)	0.9	—	—	—
	♀Black East Indian	♀♀(2)	3.8	2.8 to 4.7	—	—

TABLE XXIV

WHITE PRIMARIES

(from Jaap, 1933a)

	Genotype	Phenotype
RR	WW	runner (includes white primaries)
	Ww	runner (,, ,, ,,)
	ww	runner (,, ,, ,,)
Rr	WW	white primaries
	Ww	white primaries
	ww	white primaries
rr	WW	coloured primaries
	Ww	coloured primaries
	ww	white primaries

TABLE XXV

TO SHOW PHENOTYPIC EFFECTS OF THE RUNNER FACTOR IN THE HOMOZYGOUS
AND HETEROZYGOUS CONDITIONS

	Parents		Phenotypes of Progeny					
Mating Number	Breeds	Geno-types	Full Runner Pattern (RR)	Neck Stripe and White Primaries (Rr)	Neck Stripe only (Rr)	White Primaries only (Rr)	No White Markings (Rr)	White with "Ghost" Runner Pattern (cc)
65	♂Rouen ♀"Runner"*	rreeCc RReeCc	—	3	5	—	7	2 (Rr)
66	♂"Runner" ♀Cayuga	RReeCc rrEECC	—	?	?	6**	6**	—
64	♂"Runner" ♀"Runner"	RReeCc RReeCc	28	—	—	—	—	11 (RR)
67	♂"Runner" ♀"Runner"	RReeCc RReeCC	27	—	—	—	—	—

* All the "Runner" birds in these matings were derived from the F_2 generation of a Khaki Camp-bell – White Campbell cross.

** White bib present with possible masking of Neck Stripe.

TABLE XXVI

TO DEMONSTRATE THE INHERITANCE OF RECESSIVE WHITE IN THE DUCK

Mating Number	Breeds	Genotypes	Coloured (C)	White (c)
	Parents		*Phenotypes of Progeny*	
47	♂Khaki Campbell	CC	52	—
	♀White Campbell	cc		
50	♂White Campbell	cc	24	—
	♀Khaki Campbell	CC		
16	♂Black Cayuga	CC	9	—
	♀White Campbell	cc		
15	♂Welsh Harlequin	CC	39	—
	♀White Campbell	cc		
		Total	124	—
		Expected	124	—
53	♂F_1 – mating 47	Cc	33	33
	♀White Campbell	cc		
28	♂F_1 – mating 16	Cc	58	48
	♀White Campbell	cc		
		Total	91	81
		Expected	86	86
	$\chi^2 = 0.581, .3 < P < .5$			
54	♂F_1 – mating 47	Cc	128	44
	♀F_1 – mating 47	Cc		
27	♂F_1 – mating 16	Cc	209	69
	♀F_1 – mating 16	Cc		
64	♂F_1 – mating 54	Cc	28	11
	♀F_1 – mating 54	Cc		
65	♂Rouen (Cc)	Cc	18	2
	♀F_1 – mating 54	Cc		
		Total	383	126
		Expected	381.75	127.25
	$\chi^2 = 0.016, P = .9$			
55 & 56	♂F_1 – mating 54 (White)	cc	—	288
	♀F_1 – mating 54 (White)	cc		
		Expected		288
67	♂F_1 – mating 54	Cc	27	—
	♀F_1 – mating 54	CC		
66	♂F_1 – mating 54	Cc	12	—
	♀Black Cayuga	CC	12	—
		Total	39	—
		Expected	39	—

TABLE XXVII

SHOWING RESULTS OF MATINGS CARRIED OUT TO DETECT COMPLEMENTARY FORMS
OF WHITE IN THE DUCK

Mating Number	Parent Breeds	Coloured Progeny		White Progeny	
		Males	Females	Males	Females
57	♂Aylesbury ♀White Campbell	—	—	109	97
58	♂Aylesbury ♀White Campbell	—	—	48	44
	Total	—	—	157	141
59	♂White Decoy ♀White Indian Runner	—	—	—	1
60	♂White Indian Runner ♀White Decoy	—	—	29	28
	Total	—	—	29	29
61	♂White Decoy ♀White Campbell	—	—	5	5

TABLE XXVIII

SHOWING EVIDENCE OF INDEPENDENT ASSORTMENT BETWEEN RECESSIVE WHITE
(c) AND EXTENDED BLACK (E)

Parents			Phenotypes of Progeny		
Mating Number	Breeds	Genotypes	CE	Ce	cE and ce (c epistatic to E and e)
16	♂Black Cayuga ♀White Campbell	EECC eecc	9	—	—
		Expected	9	—	—
27	♂F_1 – mating 16 ♀F_1 – mating 16	EeCc EeCc	100	42	45
		Expected (3 : 1) χ^2(CE, Ce) = 1.587	106.5 0.2 < P < 0.3	35.5	
		Expected (3 : 1) χ^2(C, c) = 0.0873	140.25 0.7 < P < 0.8		46.75
28	♂F_1 – mating 16 ♀White Campbell	EeCc eecc	30	23	48
		Expected (1 : 1) χ^2(CE, Ce) = 0.924	26.5 0.3 < P < 0.5	26.5	
		Expected (1 : 1) χ^2(C, c) = 0.248	50.5 0.5 < P < 0.7		50.5

APPENDIX

Dominant	Wild Type (+)	Recessive	Lower Recessive
restricted (M^R)	mallard (M)	dusky (m^d)	—
	dark phase (Li)	light phase (li)	harlequin phase (li^h)
extended black (E)	non-extended black (e)	—	—
blue dilution (G)	non-blue dilution (g)	—	—
—	non-brown dilution (D)	brown dilution (d)	—
—	non-buff dilution (Bu)	buff dilution (bu)	—
dominant bib (S)	non-bib (s)	—	—
—	non-bib (B)	recessive bib (b)	—
—	coloured primaries (W)	white primaries (w)	—
runner (R)	non-runner (r)	—	—
—	coloured plumage (C)	white plumage (c)	—
white skin and bill colour (Y)	yellow skin and bill colour (y) (Rendel, 1940)	—	—
—	green egg colour (P) (Mazing, 1933)	white egg colour (p)	—
crested (Cr)	non-crested (cr) (Rust, 1932)	—	—
	normal (A) (Dyrendahl, 1958)	hereditary tremour (a)	—

N.B. Neither RENDEL's paper nor the English Abstract of the report of MAZING's work quote any symbols. The ones used in the above table have been suggested by the author.

Linkage Groups
No. 1 – autosome – extended black (E) and dominant bib (S)
No. 2 – sex chromosome – brown dilution (d) and buff dilution (bu).

REFERENCES

APPLEYARD, R., (1949). *Ducks*. Poultry World Ltd., London.

ASMUNDSON, V. S., (1940). A recessive slate plumage colour in turkeys. *J. Heredity* **31**: 215.

ASMUNDSON, V. S., (1945). A triple-allele series and plumage colour in turkeys. *Genetics* **30**: 305–322.

BATESON, W., and R. C. PUNNETT, (1906). Experimental studies in the physiology of heredity. *Poultry Repts. Evol. Comm. Roy. Soc.* **III**: 11–30.

CASTLE, W. E., (1916). Further studies on piebald rats and selection. *Carnegie Inst. Wash. Publ.* No. **241.**

CASTLE, W. E., (1919). Piebald rats and the theory of genes. *Proc. Nat. Acad. Sci. Wash.* **5**: 126–130.

CASTLE, W. E., (1951). Variation in the hooded pattern of rats, and a new allele of hooded. *Genetics* **36**: 254–266.

CASTLE, W. E., and J. C. PHILLIPS, (1914). Piebald rats and selection. *Carnegie Inst. Wash. Publ.* No. **195.**

CREW, F. A. E., and S. S. MUNRO, (1938). Gynandromorphism and lateral asymmetry in birds. *Proc. Roy. Soc. (Edin.)* **58**: 114–134.

DELACOUR, J., (1956). *Waterfowl of the World*. Volume **II**. Country Life, London.

DU SHANE, G. P., (1944). The embryology of vertebrate pigment cells. Part II. Birds. *Quart. Rev. Biol.* **19**: 98–117.

DYRENDAHL, S., (1958). Hereditary tremor in ducks. *J. Heredity* **49**: 214–216.

FINN, F., (1913). Some spontaneous variations in Mallard and Muscovy ducks. *Avicultural Magazine* **4**: 82–87.

FOX, H. M., (1955). The colours of animals. *Endeavour* **14**: 40–47.

GHIGI, A., (1924). On the inheritance of colour in the guinea fowl. *Proc. 2nd. World's Poultry Congress (Barcelona)*: 18.

GHIGI, A. (1927). *Monografia della galline di faraone*. Publicazioni della stazione sperimentale di pollicoltura di Rovigo.

GHIGI, A., (1936). *Galline di Faraone e Tacchini*. Hoepli, Milan.

GHIGI, A., and A. M. TAIBEL, (1927). Investigations concerning the heredity of colour in ducks. *Third World's Poultry Congress Report*: 147–149.

GOODALE, H. D. (1911). Studies on hybrid ducks. *J. Exp. Zool.* **10**: 241.

HARRISON, J. M., (1959). Analogous variation in Mallard and its possible significance. *Bull. B.O.C.* **79**: 22–25.

HOLLANDER, W. F., (1944). Mosaic effects in domestic birds. *Quart. Rev. Biol.* **19**: 285–307.

HOLLANDER. W. F. and P. L. WALTHER, (1962), Recessive „Lavender" in the Muscovy duck. *J. Heredity* **53:** 81.

HUNTER, J. R., (1939). A light mutant of the Mallard duck. *J. Heredity* **30:** 546–548.

HUNTER, J. R., (1950). *Magazine of Ducks and Geese,* **1** (4): 12.

HUTT, F. B., (1949). *Genetics of the Fowl.* McGraw Hill Book Co., New York.

HUTT, F. B., (1951). Snow-white down in the chick. *J. Heredity* **42:** 117–120.

HUTT, F. B. & C. D. MUELLER, (1942). Sex-linked albinism in the turkey. *J. Heredity* **33:** 69–77.

JAAP, R. G., (1933a). Inheritance of white spotting in ducks. *Poultry Sci.* **12:** 233–241.

JAAP, R. G., (1933b). Light phase Mallard ducks. *J. Heredity* **24:** 467.

JAAP, R. G., (1934). Alleles of the Mallard plumage pattern in ducks. *Genetics* **19:** 310.

JAAP, R. G., & W. F. HOLLANDER, (1954). Wild type as standard in poultry genetics. *Poultry Sci.* **33:** 94–100.

JAAP, R. G. & T. T. MILBY, (1944). Comparative genetics of blue plumage in poultry. *Poultry Sci.* **23:** 3–8.

JEROME, F. N., (1953). Colour inheritance in geese and its application to goose breeding. *Poultry Sci.* **32:** 159.

KAGELMANN, G., (1951). Studien über farbfelderung, Zeichnung und, farbüng der wild und hausenten. *Zool. Jb., Physiol.* **62:** 513–630.

KORTRIGHT, F. H., (1943). The Ducks, Geese and Swans of North America. American Wildlife Institute.

KRIZENECKY, J., (1961). Snow-white down in ducks. *J. Heredity* **52:** 237–239.

LAMON, H. M. & R. R. SLOCUM, (1922). Ducks and Geese. Orange Judd Pub. Co., New York.

LAMOREUX, W. F., & F. B. HUTT, (1942). Variations in the down colour of White Leghorn chicks and their economic significance. *J. Agric. Res.* **64:** 193–205.

LEVI, W. M., (1957). The Pigeon. Levi Publ. Co. Inc.

LIPPINCOTT, W. A., (1918). The case of the Blue Andalusian. *Amer. Nat.* **52:** 95–115.

LIPPINCOTT, W. A., (1923). Genes for the extension of black pigment in the chicken. *Amer. Nat.* **57:** 284–287.

LLOYD-JONES, O., (1915). Studies on inheritance in Pigeons. II A microscopical and chemical study of feather pigments. *J. Exp. Zool.* **18:** 453–509.

MacCURDY, HANSFORD, & W. E. CASTLE, (1907). Selection and crossbreeding in relation to the inheritance of coat pigments and coat patterns in rats and guinea pigs. *Carnegie Inst. Wash. Publ.* No. **40.**

McILHENNY, E. A., (1941). Unusual plumage of domestic Mallard ducks. *J. Heredity* **32:** 18–21.

MASON, C. W., (1923). *J. Phys. Chem.* **27:** 401.

MASON, C. W., (1927). *J. Phys. Chem.* **31:** 1856.

MATTHEWS, G. V. T., (1959). British Literature on European wildfowl. *The Wildfowl Trust Tenth Annual Report* (1957–8): 162–175.

MAZING, R. A., (1933). K voprosu o proishož-denii domašnik utok. (The origin of the domestic duck. Origin of domesticated animals. Published by U.S.S.R. Academy of Sciences: 253–258. English Abstract – Anim. Breed. Abstr. (1934). 2: 149.

MILBY, T. T., (1943). A suggested method of keeping small white turkeys distinct from White Hollands. Poultry Sci. 22: 395.

MUELLER, C. D., & F. B. HUTT, (1941). Genetics of the fowl. 12. Sex-linked, imperfect albinism. J. Heredity 32: 71–80.

MUNRO, S., (1946). A Sex-linked true-breeding blue plumage colour. Poultry Sci. 25: 408.

OGILVIE, M. A., (1961). British literature on European wildfowl, 1957–60. The Wildfowl Trust Twelth Annual Report (1959–60): 157–162.

PHILLIPS, J. C., (1915). Experimental studies of hybridization among pheasants and ducks. J. Exp. Zool. 18: 69–144.

PHILLIPS, J. C., (1921). A further report on species crosses in birds. Genetics 6: 366.

PUNNETT, R. C., (1930). A sex-linked character in ducks. Nature (London) 26: 757.

PUNNETT, R. C., (1932). Note on a sex-linked character in ducks. J. Genetics 25: 191–194.

RENDEL, J. M., (1940). Note on the inheritance of yellow bill colour in ducks. J. Genet. 40: 439–440.

RIDGWAY, R., (1912). Colour standards and colour nomenclature. Published by author, Washington, D.C., 1, 115 named colours.

ROBERTSON, W. R. B., (1925). Evidence of somatic segregation of factors for plumage colour in heterozygotes in the turkey. (Abstract.) Anat. Rec. 31: 356.

ROBERTSON, W. R. B., B. B. BOHREN and D. C. WARREN, (1943). The inheritance of plumage colour in the turkey. J. Heredity, 34: 246–256.

ROGERON, G., (1903). Les Canards. J. B. Bailliere et fils, Paris.

RUST, W., (1932). Lethalfaktoren und unvollkommene Dominanz bei Haubenten Arch. Geflügelk. 6: 110–116.

SAGE, B. L., (1953). On some unusual plumage variations in the mallard. Bull. B.O.C. 73: 60–61.

SAGE, B. L., (1954a). A recent example of symmetrical albinism in the mallard. Bull. B.O.C. 74: 30.

SAGE, B. L., (1954b). On the plumage characters of an aberrant female mallard. Bull. B.O.C. 74: 74.

SAGE, B. L., (1955). Some further notes on plumage variations in the mallard. Bull. B.O.C. 75: 54–57.

SCOTT, P., (1957). A Coloured Key to the Wilfowl of the World. The Wildfowl Trust: 25.

SEREBROVSKY, A. S., (1926). Somatic segregation in the domestic fowl. J. Genetics 16: 33–42.

SOKOLAVSKAJA, I. I., (1935). Opyty po gibridizacii ptic. I. Sceplennye s polom priznaki u gibridov mež-du Cairina moschata i Anas platyrhyncha. Zool.

ž. (*Mosk.*) **14:** 481–496. English Abstract: *Anim. Breed. Abstr.* (1936) **4:** 228.

TAIBEL, A. M., (1954). Il plumaggio "grigio – ardesia – azzurastro" del' anatra muschiata (Cairina moschata domestica L.) e suo valore genetico. Creazione della varietà "Grigio – perla" per mutazione. *Ann. Sper. Agr., N.S.* **8:** 1795–1801. English Abstr. *Anim. Breed. Abstr.* (1955) **23:** 191.

VAN ALBADA, M. and A. R. KUIT, (1960). Een geslachtsgebonden verdunningsfactor voor veerkleur bij Witte Leghorns. *Genen en Phaenen* **5**: 1–9.

WALTHER, A. R., J. HAUSCHILDT and J. PRÜFER, (1932). Ein wirtschaftlich wichtiger, geschlechtsgebundener Faktor bei Enten. *Der Züchter* **4:** 18–22.

WARREN, D. C., (1940). Inheritance of pink ye in the fowl. *J. Heredity* **31:** 291–292.

WRIGHT, L., (1902). *The new book of poultry.* Cassell & Co. Ltd., London.

ACKNOWLEDGEMENTS

The suggestions and encouragement of Dr. H. TEMPERTON throughout the work are gratefully acknowledged. The author is also indebted to Mr. F. J. DUDLEY and Dr. N. K. JENKINS for reading the manuscripts and for their constructive criticisms and suggestions, and to Mr. T. O. WILSON, Dr. N. K. JENKINS and Mr. C. M. GROOM for their valuable photographic assistance. Finally, thanks are due to Mr. A. RICE for kind permission to use his photographs of the Magpie, Muscovy and Runner ducks.

Plate 1. DAY-OLD NON-DILUTED DUCKLINGS
      Left: – "restricted"
      Centre: – "mallard"
      Right: – "dusky"

Plate 2. DAY-OLD BROWN-DILUTED DUCKLINGS
    Left: – "restricted"
    Centre: – "mallard"
    Right: – "dusky" (The Khaki Campbell variety bears this genotype –
    $m^d m^d d)d$))

Plate 3. TOP ROW – NON-DILUTED DAY-OLD DUCKLINGS
      Left: – "dark phase"
      Centre: – "light phase"
      Right: – "harlequin phase"
      BOTTOM ROW – BROWN-DILUTED DAY-OLD DUCKLINGS
      Left: – "dark phase"
      Centre: – "light phase"
      Right: – "harlequin phase"

Plate 4. NON-DILUTED MATURE FEMALES
      Left: – "harlequin phase"
      Centre: – "light phase"
      Right: – "dark phase"

Plate 5. MATURE FEMALES
    (a) Pale Blue Orpington (GGEE)
    (b) Dark Blue Orpington (GgEE)   Note black flecking
    (c) Black Cayuga (ggEE)

Plate 6. ADULT PALE BLUE ORPINGTONS
    Left: – Female
    Right: – Male

Plate 7. TOP ROW – NON-EXTENDED BLACK DAY-OLDS
    (a) Khaki (ggd(d)ee)
    (b) Dark Buff (Ggd(d)ee)
    (c) Pale Buff (GGd(d)ee)
    BOTTOM ROW – EXTENDED BLACK DAY-OLDS
    (d) Black (ggD(D)EE)
    (e) Dark Blue (GgD(D)EE)
    (f) Pale Blue (GGD(D)EE)

Plate 8. TOP ROW – EXTENDED BLACK DAY-OLDS
    (a)   Black male (ggDdEE)
    (b)   Chocolate female (ggd–EE)
    (c)   Blue male (GgDdEE)
    (d)   Lilac female (Ggd–EE)
BOTTOM ROW – NON-EXTENDED BLACK DAY-OLDS
    (e)   Mallard male – "dusky" (ggDdee)
    (f)   Khaki female – "dusky" (ggd–ee)
    (g)   Blue Mallard male (GgDdee)
    (h)   Buff female (Ggd–ee)

Plate 9. MATURE FEMALES
    (a) Pale Buff Orpington (GGd–ee)
    (b) Dark Buff Orpington (Ggd–ee)
    (c) "Khaki" Orpington (ggd–ee)

Plate 10. MATURE FEMALES – EXTENDED BLACK
    (a) Lilac (Ggd–EE)
    (b) Chocolate (ggd–EE)
    (c) Blue (GgD–EE)
    (d) Black (ggD–EE)

Plate II. DAY-OLD HYBRID DUCKLINGS
    (a) Non-diluted male (ggDd)
    (b) Brown diluted female (ggd–)
    (c) Blue diluted male (GgDd)
    (d) Brown and Blue diluted female (Ggd–)

Plate 12. TOP ROW – DAY-OLD DARK BUFF DUCKLINGS
    (a) Non-diluted male (ddGgBubu)
    (b) Buff diluted female (d–Ggbu–)
    BOTTOM ROW – DAY-OLD KHAKI DUCKLINGS
    (c) Non-diluted male (ddggBubu)
    (d) Buff diluted female (d–ggbu–)

Plate 13. DOMINANT BIB – BLUE ORPINGTON DRAKE

Plate 14. BLACK EAST INDIAN DUCK, SHOWING WHITE MARKINGS WHICH DEVELOPED
IN THE SECOND SEASON

Plate 15. "PLUS SERIES" DOMINANT BIB – F₂ GENERATION DRAKE FROM BLACK EAST
INDIAN × BLUE ORPINGTON

Plate 16. F₁ GENERATION DUCK FROM KHAKI CAMPBELL × BLUE ORPINGTON

Plate 17. BLACK AND WHITE MUSCOVY DRAKE

Plate 18. MAGPIE DUCK

Plate 19. BREEDING GROUP OF FAWN AND WHITE INDIAN RUNNERS